NUTRITION IN
THE LOWER METAZOA

Proceedings of a meeting held at the University of Caen,
France, 11-13 September 1979

Editors

D. C. SMITH

Sibthorpian Professor of Rural Economy
University of Oxford, England

and

Y. TIFFON

Laboratoire Maritime
Université de Caen, France

PERGAMON PRESS

OXFORD · NEW YORK · TORONTO · SYDNEY · PARIS · FRANKFURT

U.K.	Pergamon Press Ltd., Headington Hill Hall, Oxford OX3 0BW, England
U.S.A.	Pergamon Press Inc., Maxwell House, Fairview Park, Elmsford, New York 10523, U.S.A.
CANADA	Pergamon of Canada, Suite 104, 150 Consumers Road, Willowdale, Ontario M2J 1P9, Canada
AUSTRALIA	Pergamon Press (Aust.) Pty. Ltd., P.O. Box 544, Potts Point, N.S.W. 2011, Australia
FRANCE	Pergamon Press SARL, 24 rue des Ecoles, 75240 Paris, Cedex 05, France
FEDERAL REPUBLIC OF GERMANY	Pergamon Press GmbH, 6242 Kronberg-Taunus, Pferdstrasse 1, Federal Republic of Germany

First edition 1980

British Library Cataloguing in Publication Data

Nutrition in the Lower Metazoa *(Conference),*
University of Caen, 1979
Nutrition in the lower metazoa.
1. Invertebrates - Feeding and feeds -
Congresses
I. Smith, David Cecil
II. Tiffon, Y
593   QL364   80-40081
ISBN 0-08-025904-9

In order to make this volume available as economically and as rapidly as possible the authors' typescripts have been reproduced in their original forms. This method has its typographical limitations but it is hoped that they in no way distract the reader.

Printed in Great Britain by A. Wheaton & Co. Ltd., Exeter

CONTENTS

Contents

INTRODUCTION

One of the most important early contributions to the study of nutrition in the lower metazoa was made by Mestchnikoff (although famous for his contributions to pathology, it is less known that he first had a very productive career as a zoologist). He not only discovered phagocytosis in the Turbellaria (cf. De Duve, 1969) but at the same time as Chapeaux (1893) he demonstrated that intracellular digestion begins with an acid stage in the Actinaria. Thus these phenomena had been discovered in the lower invertebrates long before the concept of lysosomes began to arise.

After Mestchnikoff (1880), the direction of research into nutrition of lower metazoa became diversified and intensified, especially into enzymological studies. It was shown that proteases such as trypsin and chymotrypsin previously known from vertebrates also existed in all Anthozoa so far examined. In Metridium senile, one of the three serine proteases is completely homologous with mammalian trypsin (Gibson & Dixon, 1969). Similarly, the acid hydrolases of Anthozoa have characteristics identical with those of higher vertebrates. Thus it is unlikely that detailed studies of the enzymes of the lower invertebrates will throw new light on general problems of nutrition.

On the other hand, the nutritional methods used by the lower invertebrates have undoubtedly been successful, and this certainly merits further study. For example, the Turbellaria have colonised marine, freshwater and terrestrial habitats, the latter including parasitic forms, both endo- and ecto-symbiotic. Their nutritional physiology therefore shows great plasticity, and the biochemical study of such adaptability would be of great value. The Sponges and Coelenterates are further examples of success. These two groups, already well represented in temperate seas, attain their fullest success in tropical oceans which are notoriously poor in nutrients. Nevertheless, tropical reefs, with their rich fauna and flora, are amongst the most productive ecosystems in the world. In the case of Coelenterates and Sponges in this ecosystem, the formation of mutualistic symbioses with algae and bacteria are of key importance to their nutritional success and ecological dominance. An additional feature in the success of lower invertebrate groups is their ability to absorb and utilise both particulate and dissolved organic matter from their aqueous environments.

The lower metazoa are therefore interesting because of their ecological success, and because of both their whole organism and cellular physiology. The study of the nutritive mechanisms they employ may enable us to pose, and perhaps also solve, problems of a more general nature in understanding cell function.

It was with this background that a three day discussion meeting on 'Nutrition in the lower metazoa' was held at the University of Caen, Normandy, 11-13 September 1979. The papers delivered at that meeting form the contributions to this volume. They illustrate the value of bringing together, around a common theme, different people working with different organisms, and in a constructive attempt to redefine a number of questions, some of them perhaps new.

D.C. Smith   Y. Tiffon

Chapeaux, M. (1893). Recherches sur la digestion des Coelenteres. Arch. Zool.
   Exp. gen., 3, 279-289.
De Duve, C. (1969). The lysosomes in retrospect. In J.T. Dingle and H.B. Fell(Eds)
   Lysosomes in Biology and Pathology, Vol.1, North Holland/American Elsevier,
   Amsterdam, London, New York, pp. 3-40.
Gibson, D. & G.H. Dixon (1969). Chymotrypsin like protease from the sea
   anemone Metridium senile. Nature, 222, 753-756.
Metschnikoff, E. (1880). Uber die intracellulare Verdauung bei Coelenteraten.
   Zool. Anz., 3, 56, 261-263.

PHAGOCYTOSIS IN *POLYCELIS TENUIS*

I. D. BOWEN

Department of Zoology, University College Cardiff, Cardiff,
Wales, U.K.

ABSTRACT

On feeding, occlusion of the intestinal lumen has been observed in this organism.
Columnar phagocytes from opposing sides come together and interdigitate apically,
forcing food particles, consisting of semi-digested tissues and cells, into the
gastrodermis. In some instances, the phagocytosis of intact cells can be seen.
Using horse-radish peroxidase as a marker, phagocytosed material can be histochemically detected a few minutes after commencement of feeding. Mixing of phagosomes
and lysosomes occurs between 30 minutes and 2 hours after feeding, and digestion
continues for up to 5 hours, as demonstrated by the presence of acid phosphatase.
The fine structure of the gastrodermis is reviewed and the relationships between
phagocytosis, enzyme production and intracellular digestion investigated. Primary
lysosomes containing acid phosphatase appear to be formed at the Golgi apparatus,
basally in the gastrodermal phagocytes, and then fuse with the incoming phagosomes,
forming digestive vacuoles or secondary lysosomes. The digestive vacuoles finally
end up as residual bodies. Autophagy and cell death are demonstrated and their
role in nutrition is discussed.

KEYWORDS

Fine structure; gastrodermis; phagocytosis; horse-radish peroxidase; acid phosphatase; autophagy; starvation; cell death.

INTRODUCTION

Recently the phenomenon of phagocytosis has been allied with the lysosome concept.
C. de Duve and others have demonstrated that lyscsomal enzymes are largely involved
with the degradation of metabolites and that compounds broken down in the lysosomes
have in general been taken up from the environment by pinocytosis and phagocytosis.
Autophagy, the isolation and eventual digestion of material derived from the cell's
own cytoplasm, can be considered as a special kind of phagocytosis.

Pinocytosis and phagocytosis were known to cytologists a considerable time before
the lysosome concept was formulated (Metchnikoff, 1892, 1901). The term phagocytosis is usually applied to vacuolar uptake of solid particles by cells ; Straus
(1959), however, employs the term 'phagosome' to describe particles which may inc-

1

lude material ingested by pinocytosis or phagocytosis. Straus (1959, 1962) has
developed an elegant technique for following the fate of phagosomes within cells
using the enzymatic label horse-radish peroxidase. This technique may be coupled
with the cytochemical demonstration of acid phosphatase in the same tissue section
(Straus, 1964, 1967a,1967b). In this way the digestion and fate of phagocytosed
material can be followed and its involvement with acid phosphatase-positive lysosomes can be studied. The relationships between lysosomes and phagosomes has been
amply reviewed by Straus (1967b).

The process of phagocytosis has been extensively studied in planaria at the level
of the optical microscope (Willier, Hyman and Rifenburgh, 1925; Hyman, 1951; Osborne
and Miller, 1962; Horne and Darlington, 1967). The information obtained in these
earlier studies has been reviewed by Jennings (1968). In this investigation planaria were fed with horse-radish peroxidase as described by Osborn and Miller (1962)
in order to demonstrate phagocytosis. This method was also combined with the acid
phosphatase technique as described by Straus (1964, 1967a) in rat kidney and liver.
This combination of techniques has permitted a physiological interpretation of the
relationship between gastrodermal phagosomes and lysosomes. The uptake of food
material and electron opaque thorium oxide by phagocytosis has been confirmed at
electron microscope level.

There have been a number of morphological and histochemical studies on the planarian intestine (Willier, Hyman and Rifenburgh, 1925; Hauser, 1956; Rosenbaum and
Rolon, 1960; Jennings, 1957, 1962; Rose and Shostak, 1968) covering several species.
The first morphological description relating to Polycelis tenuis is that of Iijima
(1884). Fine structural studies on the planarian gut are surprisingly few, perhaps
due to fixation difficulties; Pedersen (1961) made some observations on the gastrodermal phagocytes of Dugesia tigrina and Ishii (1965) described two gastrodermal
cells in the freshwater species Bdellocephala brunnea. Detailed studies of the
fine structure of the pharynx and intestine of Polycelis tenuis were presented by
Bowen and Ryder (1973) and Bowen, Ryder and Thompson (1974) respectively; aspects
of which are reviewed here.

Several histochemical studies (Rosenbaum and Rolon, 1960; Jennings, 1962; Osborne
and Miller, 1963) suggest an intracellular digestive role for various hydrolytic
enzymes in the planarian gut. Electron cytochemical studies on the fine structural
distribution of acid phosphatase (sodium- β-glycerophosphatase) in the digestive
epithelial cells have been undertaken by Bowen, Ryder and Thompson (1974). Ryder
and Bowen (1974, 1975) also describe the electron cytochemical localization of
alkaline phosphatase and several acid phosphatases in the gastrodermis, and other
tissues from Polycelis tenuis. Bowen, Ryder and Dark (1976) and Bowen and Ryder
(1976) describe the fine structural localization of p-nitrophenyl phosphatase in
starving planaria. Bowen and Ryder (1974) also report cell deletion in normal planarian gastrodermis.

              RESULTS

Intestinal Occlusion

On feeding, occlusion of the intestinal lumen has been observed in Polycelis tenuis
(Figs. 1, 2 and 3). Columnar phagocytes from opposing sides come together and
interdigitate apically, forcing food particles consisting of semi-digested tissues
and cells into the gastrodermal phagocytes. In some instances, the phagocytosis
of intact cells can be seen (Fig. 6). The extensive and complicated interlocking
of cells seen during occlusion makes it unlikely that the phenomenon represents a
fixation artefact. The closure observed is most complete near pharyngeal insertion

but does extend throughout the intestine to a varying degree. Similar profiles are

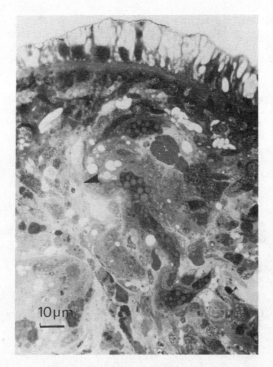

Fig. 1. Occlusion of intestinal lumen (arrow).

produced by differing preservative techniques e.g. Carnoy's fixative, 3% buffered
glutaraldehyde, and immediate freezing in liquid nitrogen; again suggesting that
the observed closure is not a fixation induced artefact.

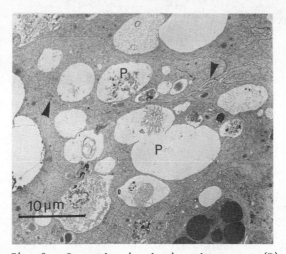

Fig. 2. Gastrodermis showing phagosomes (P)
and closure of lumen (arrow)

Occlusion of the intestinal lumen after feeding is a phenomenon which deserves closer attention. Early papers by Metchnikoff (1901) and Saint-Hilaire (1910) and
more recently Hauser (1956) refer to a degree of gastrodermal reorganization after
feeding. In starved animals an extensive lumen can be seen (Ishii, 1965). The
closure of the lumen after feeding is spectacular and virtually complete. Although
there is no evidence of any loss of cellular material as suggested by Hauser (1956),
luminal occlusion could be a way of achieving extensive and more or less instantaneous phagocytosis. Experimental evidence employing horse-radish peroxidase indicates a rapid uptake of material in a matter of minutes. Contraction of the gut by
surrounding muscular elements could produce the effect of a press, forcing food
from the closing lumen into the opposing phagocytic cells. Complete luminal occlusion shown here and also demonstrated by Ishii (1965) in fed Bdellocephala is not
consistent with the amoeboid-type of activity normally attributed to these cells.
The gastrodermal phagocytes bear numerous marginal microvilli (Figs. 2 and 3) these
are coated with a fine glycocalyx. These microvilli are often involved in the
process of interdigitation during occlusion immediately after feeding.

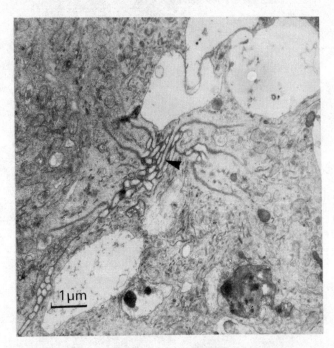

Fig. 3. Interdigitating microvilli (arrow) bordering
         occluded intestinal lumen.

Uptake Studies

Uptake studies were carried out on Polycelis starved for 7 days or more to ensure
immediate feeding and to help clear the gut of previously contained material. The
animals were fed on a paste made up of crushed earthworm and horse-radish peroxidase. Planaria were removed at 5, 10, 20, 30 minute and 1, 2, 3, 4 and 5 hour
intervals after commencement of feeding and killed in Baker's formalin. Fixed
animals were quenched in liquid nitrogen and sectioned in a cryostat. Cryosections
were air dried on coverslips and reacted for both acid phosphatase (Burstone, 1962)
and peroxidase (Straus, 1967a) (benzidine test modified after Gomori 1952).

The electron dense properties of thorium have previously been made use of by Padawer
(1962). Animals were fed on a paste of thorium oxide and chopped earthworm or liver.
(0.5 gm. thorium oxide/5 gms. food). Neutral red can be added as a useful feeding
indicator. Animals were also fed on chopped liver alone.

Results of the histochemical azo dye test for intrinsic acid phosphatase activity
showed a positive reaction in the gastrodermis. This activity is enhanced on feeding. Localization takes the form of small particles or vacuoles containing red
reaction product (using red violet LB salt as coupler). The extent of positive
reaction increased up to 10 minutes after feeding and persisted at this high level
for several hours. Introduced peroxidase had already been phagocytosed by columnar
cells as early as 5 minutes after commencement of feeding. The peroxidase reaction
product appeared initially as a blue colour within apical particles or vacuoles.
Between 30 minutes and 2 hours after feeding a polarization of the blue peroxidase
vacuoles and the red acid phosphatase vacuoles was observable, the former occurring
apically, the latter mainly but not entirely basal. During this period some of the
particles could be seen to fuse producing larger purple coloured vacuoles (Fig. 4),
presumably digestive vacuoles or secondary lysosomes. The purple vacuoles predom-
inated between 2 to 5 hours after feeding and subsequent to this time no separate

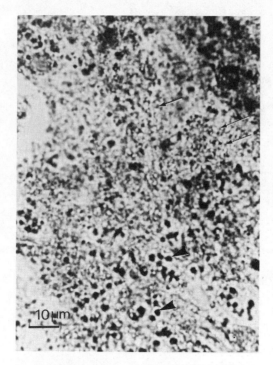

Fig. 4. Uptake of horse-radish peroxidase (small arrows)
and fusion with acid phosphatase positive vacuoles
to form digestive vacuoles (large arrows).

blue reaction for peroxidase was visible. Separate red particles indicative of acid
phosphatase activity persist at a basal level. No intrinsic peroxidase activity
could be detected in the gastrodermis using the benzidine test. Whilst phagocytosis
was common in the intestine no phagocytosis was apparent in the inner epithelium of
the pharynx at any time.

Electron dense particles of thorium oxide were found in food vacuoles in the apical
portions of the phagocytic cell between 15 and 30 minutes after feeding. By this
time thorium oxide was found to occur throughout the intestine and in cells at the
extreme ends of both anterior and posterior rami of the intestine. Unfortunately
difficulty was encountered in obtaining colloidal thorium dioxide and the deposits
obtained were somewhat dense and difficult to section (Fig. 5).

Although some extracellular digestion does occur in Polycelis as shown by the
occurrence of lysed cells and food material in the pharynx (Bowen and Ryder, 1973),
whole cells are commonly phagocytosed. Figure 6, shows intact liver cells incorporated into a phagocytic cell 24 hours after feeding. At this stage the liver
cell contents are still largely intact although noticeably darker showing a higher
affinity for counterstain. Liver cells may be phagocytosed in large numbers so that
the gastrodermis appears like a mosaic. This may contribute to the overall occlusion of the gut lumen.

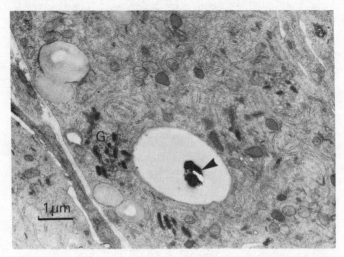

Fig. 5. Thorium oxide (arrow) in phagosome near Golgi (G).

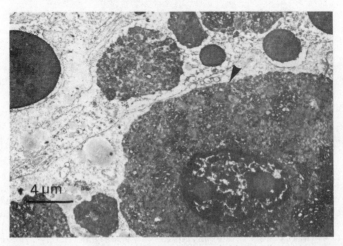

Fig. 6. Uptake of whole rat liver cell (arrow) by phagocyte.

Fine Structure and the Electron Cytochemical Demonstration of Acid Phosphatase.

The fine structure of Polycelis intestine has already been described (Bowen, Ryder and Thompson, 1974) and it is only necessary to emphasise certain points here. Two cell types can be found in the gastrodermis, these correspond to the Minotian gland cells described by Pedersen (1961) and called granular club cells by Ishii (1965), and the phagocytic columnar cells. The Minotian gland cells are also referred to as sphere cells in some papers. The gland cells and the phagocytic cells rest on a much folded basement membrane (Fig. 7) which separates them from the surrounding muscle and parenchymal cells. The latter, however, often maintain an intimate association with the basement membrane and may well have a role to play in the transfer of soluble food material from the gut. Neoblasts, which probably replace senescent intestinal cells are also quite common in this area (Fig. 7). The phagocytic or columnar cells form the major part of the intestine and are of variable shape and capacity. The denser cells probably represent the more senescent cells which are replaced by a process of cell deletion (Bowen and Ryder, 1974) involving the release of free acid phosphatase. This deletion process is accentuated by starvation (Fig. 13).

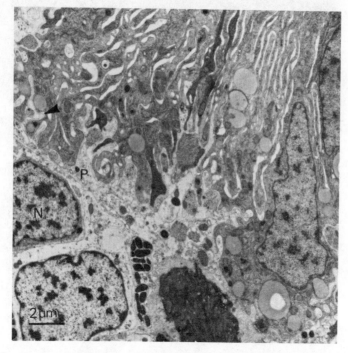

Fig. 7.   Infolded basement membrane (arrow) of phagocyte near parenchymal cell (P) and neoblast (N).

One of the most important organelles within this cell is the Golgi apparatus (Fig. 5). It is diffuse and is found in all regions of the cell but more often basally. The Golgi appears to be responsible for primary lysosome formation and also for the synthesis of multivesicular bodies. The Golgi itself frequently exhibits acid phosphatase activity (β-glycerophosphatase and p-nitrophenyl phosphatase) the distal saccules and Golgi vacuoles often showing a positive reaction (Fig. 8). A Golgi associated area of smooth membrane (defined as GERL by Novikoff, 1976) also contains acid phosphatase activity. Acid phosphatase positive Golgi vacuoles in the cell

appear to represent primary lysosomes and are morphologically identical to the dense
bodies which occur around the Golgi. After feeding these primary lysosomes or dense
bodies fuse with incoming phagosomes to produce larger digestive vacuoles or secondary lysosomes rich in acid phosphatase activity (Fig. 8). Digestion appears to

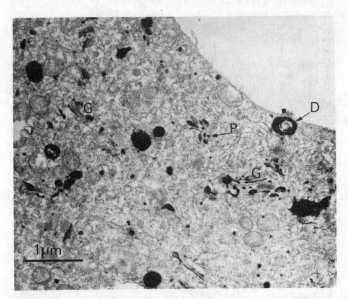

Fig. 8. Acid phosphatase activity in Golgi (G), primary
lysosomes (P) and digestive vacuoles (D).

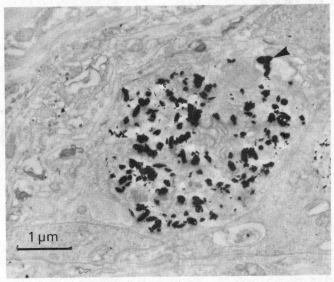

Fig. 9. Telolysosome or residual body showing residual
acid phosphatase activity (arrow)

continue within these secondary lysosomes for many hours and residual acid phosphatase can be detected within some vacuoles 24 hours after feeding (Fig. 9). These
residual bodies or telolysosomes characteristically contain many whorls or broken
membranes. The secondary lysosome population is a dynamic one and a considerable
amount of mixing occurs with digestive vacuoles fusing together to produce larger
bodies.

Autophagy

During starvation the phagocytic cells appear to digest pockets of their own cytoplasm. Autophagic vacuoles can be seen containing remains of mitochondria and other
debris (Fig. 10). These probably form under normal conditions but are morpholog-

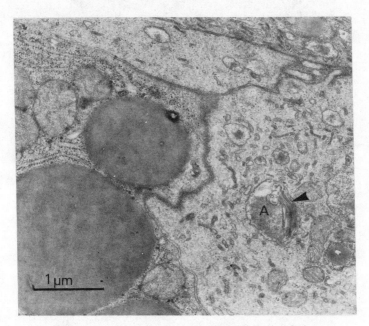

Fig. 10. Autophagic vacuole (A) containing
disintegrating mitochondrion (arrow).

ically and histochemically indistinguishable from heterophagic lysosomes since they
also acquire acid phosphatase. Bowen (1968) has demonstrated similar autophagos-
omes and autolysosomes in the digestive caeca of starved locusts. Autophagy is now
generally recognised as exerting a controlling or economic influence on the turnover of sub-cellular organelles (Ericsson, 1969). Autophagy becomes accentuated
under conditions of stress such as starvation where cells have to draw more on their
own resources. Autophagy thus plays a vital role in the nutrition of starving cells
which have no outside source of food.

Cell Deletion

Bowen and Ryder (1974) described a low level of cell deletion in normal adult planarian tissues, including the gastrodermis. In the gut lysing cells appear to be more

densely stained than normal cells. Detailed examination of such cells reveals that
the endoplasmic reticulum is vesiculated. Broken membranes can also be seen as further morphological evidence of lysis. Typically in very dense cells membranes and
vacuoles become lined with small droplets or globules of dense lipid-like material
(Fig. 11). Such cells always show an abundance of acid phosphatase activity free

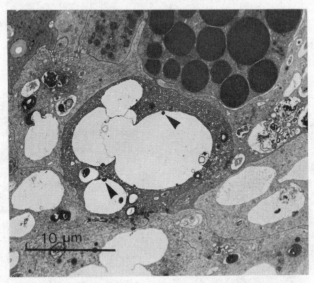

Fig. 11.   Senescent phagocyte containing dense
           lipid-like droplets (arrow).

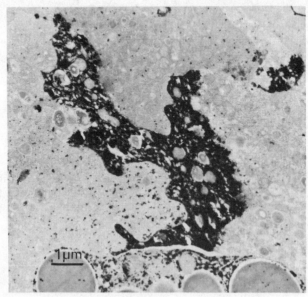

Fig. 12.   Free acid phosphatase activity in lysing
           phagocyte (uncounterstained).

in the cytoplasm (Fig. 12). As in the gland cells free acid phosphatase activity
first appears on the ribosomal side of the endoplasmic reticulum (ie. extra-cisternal) and not at its usual cisternal site. Bowen and Ryder (1974) have argued that
this extra-cisternal acid phosphatase represents newly formed or nascent enzyme
produced in order to mediate cell death.

It is assumed that evidence of progressive morphological disintegration accompanied
by increasing nascent hydrolase activity may be taken as an indication of eventual
cell death. A similar situation has been found in the gut of the slug Arion hortensis (Bowen and Davies, 1971).

Cell deletion in the normal adult planarian means that the majority of cells in the
organism are being renewed. Undifferentiated cells or neoblasts in the adult animal
are undergoing a fairly constant rate of mitosis. This cell production must be
counterbalanced by cell loss since the animal is not growing. The loss appears to
take place largely by a programmed cell lysis. Similar instances of cell death have
been described by Jones and Bowen (1979, 1980) in embryonic and newly hatched slugs
and also by Bowen and Lewis (1979, 1980) in normal and involuting mouse thymus.
Kerr and colleagues (1972, 1973) have maintained that controlled cell death or apoptosis is a phenomenon of general significance. The biological and physiological
significance of cell death is currently being reviewed by Bowen and Lockshin.

In planaria during starvation there is a distinct sequence of events; four peaks of
acid phosphatase activity can be resolved in whole animal homogenates. Three of
these have been shown to be associated with changes in the gastrodermis. The first
is associated with the immediate phagocytic response of the columnar cells; the
second (after 6 - 7 days) with increased autophagy in most cell types and the third
peak (after 14 - 15 days) is contributed to largely by the lysis and deletion of
intestinal cells (Fig. 13). Cell lysis and deletion is of relevance to nutrition
because of its conservative and economic potential. The products of lysis can be
made available to surrounding tissues.

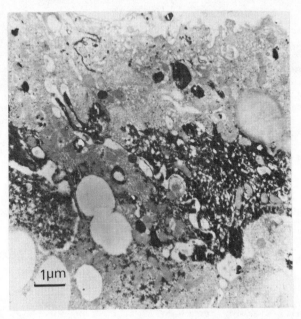

Fig. 13.   Free acid phosphatase activity in starved
           gastrodermis (uncounterstained).

The factors which cause cells to lyse in planaria are largely unknown. It is possible that each cell type has a genetically programmed life-span which is short compared with the animals as whole. Clearly the lysis of entire cells followed by the
re-cycling of cellular material may, like autophagy, be a crucial factor in the ability of planaria to survive starvation.

                    DISCUSSION AND CONCLUSIONS

Phagocytosis observed by many workers in several species of planaria (Arnold, 1909;
Willier, Hyman and Rifenburgh, 1925; Kelley, 1931; Hyman, 1951; Jennings, 1957,
1959, 1962) is confirmed here at the electron microscope level. A close correlation
has been demonstrated between phagocytosis, enzyme production and intracellular digestion. The histochemical and cytochemical results have shown a sequence of events
involving fusion of food vacuoles or phagosomes with acid phosphatase positive vacuoles or lysosomes, giving rise to digestive vacuoles or secondary lysosomes. The
fine structural studies indicate that vacuolar acid phosphatase is packaged by the
Golgi apparatus.

The time-table of digestive events presented here agrees broadly with results achieved in other cytochemical investigations (Jennings, 1957, 1962; Rosenbaum and Rolon,
1960; Osborne and Miller, 1962, 1963; Bowen, Coakley and James, 1979). Fusion of
phagosomes and lysosomes will presumably be followed by a period of digestion, resulting ultimately in the hydrolysis and absorption of the phagocytosed material.

It is not clear what factors prompt or control the rate of phagocytosis. Certain
cells have been shown to be selective in the material they will pinocytose and phagocytose. Chapman-Andresen and Holter (1955) demonstrated that the amoeba (Chaos
chaos), which does not normally pinocytose glucose could be induced to do so in the
presence of bovine serum albumin. Ryder and Bowen, (1977a, 1977b) studying aspects
of phagocytosis in the slug foot showed that epithelial cells could be induced to
endocytose materials not normally taken up when presented with ribonuclease or bovine
serum albumin.

The fate of absorbed food material has not been clearly demonstrated. The fixed
parenchymal cells, however, seem to be good candidates as a possible storage and
transfer system. The parenchymal cells occur in a very close contact with the gastrodermis and permeate throughout the animal obtaining contact with all cell types.
They may have a function similar to glial cells which they resemble. Indirect evidence of fixed parenchymal cell involvement in storage and transport of food materials
or products comes from refeeding experiments (Dark, 1978; Dark and Bowen, work in
progress). After prolonged starvation over 20 weeks the first cells to recover
exhausted lipid and glycogen stores are the fixed parenchymal cells.

                    ACKNOWLEDGEMENTS

I am very grateful to Mr. T.A. Ryder, Mr. J.A. Thompson and Mrs. Carole Winters for
technical assistance. I also wish to thank S.R.C. for the financial support of much
of the work reviewed here.

                    REFERENCES

Arnold, G. (1909). Intracellular and general digestive processes in Planaria.
    Quart. J. Micr. Sci., 54, 207-220.
Bowen, I.D. (1968). Electron cytochemical studies on autophagy in the gut epithel-
    ial cells of the locust, Schistocerca gregaria. Histochem. J., I, 141-151.

Bowen, I.D. and P. Davies (1971). The fine structural distribution of acid phosphatase in the digestive gland of Arion hortensis (Fer.). Protoplasma, 73, 73-81.

Bowen, I.D., and T.A. Ryder (1973). The fine structure of the planarian Polycelis tenuis Iijima. I. The Pharynx. Protoplasma, 78, 223-241.

Bowen, I.D., and T.A. Ryder (1974). Cell autolysis and deletion in the planarian Polycelis tenuis Iijima. Cell Tiss. Res., 154, 265-274.

Bowen, I.D., T.A. Ryder, and J.A. Thompson (1974). The fine structure of the planarian Polycelis tenuis Iijima. II. The intestine and gastrodermal phagocytosis. Protoplasma, 78, 1-17.

Bowen, I.D., and T.A. Ryder (1976). Use of the p-nitrophenyl phosphate method for the demonstration of acid phosphatase during starvation and cell autolysis in the planarian Polycelis tenuis Iijima. Histochem. J., 8, 319-329.

Bowen, I.D., T.A. Ryder, and C. Dark (1976). The effects of starvation on the planarian worm Polycelis tenuis. Cell Tiss. Res., 196, 196-209.

Bowen, I.D., W.T. Coakley, and C.J. James (1979). The digestion of Saccharomyces cerevisiae by Acanthamoeba castellanii. Protoplasma, 98, 63-71.

Bowen, I.D., and G.H.J. Lewis (1979). A method for the simultaneous estimation of mitosis and cell death. Suppl. Proc. Roy. Microsc. Soc., 14, A4.

Bowen, I.D., and G.H.J. Lewis (1980). Acid phosphatase and cell death in mouse thymus. Cell Tiss. Res. (in press).

Burstone, M.S. (1962). Enzyme histochemistry and its application to the study of neoplasms. Academic Press, New York and London.

Chapman-Andresen, C., and H. Holter (1955). Studies on the ingestion of ^{14}C glucose by pinocytosis in the amoeba Chaos chaos. Exp. Cell Res. Suppl., 3, 52-63.

Dark, C. (1978). The effects of long-term starvation and refeeding on differentiation in the planarian Polycelis tenuis Iijima. Ph.D. Thesis, University of Wales.

Ericsson, J.L.E. (1969). Mechanisms of cellular autophagy. In J.T. Dingle and H.B. Fell (Eds.), Lysosomes in Biology and Pathology, Vol. 2. North Holland Publishing Co., Amsterdam.

Gomori, G. (1952). Microscopic Histochemistry. University of Chicago Press, Chicago, Illinois.

Hauser, P.J. (1954). Histologische Umbauvorgange in Planariendarm bei der Nahrungsaufnahme. Mikroskopie, 11, 20-31.

Horne, M.K., and J.T. Darlington (1967). Uptake and intracellular digestion of ferritin in the planarian Phagocata gracilis (Haldeman). T. Amer. Microsc., 86, 268-273.

Hyman, L.H. (1951). The Invertebrates, Vol. 2. McGraw-Hill, New York.

Iijima, I. (1884). Untersuchungen uber den Bau und die Entwicklungsgeschichte der Susswasser-Dendrocoelen (Tricladen). S. f. wiss. Zoologie, 40, 359-464.

Ishii, S. (1965). Electron microscopic observation on the planarian tissues, II. The intestine. Fukushima J. Med. Sci., 12, 67-87.

Jennings, J.B. (1957). Studies on feeding, digestion and food storage in free-living flatworms (Platyhelminthes : Turbellaria). Biol. Bull., 112, 63-80.

Jennings, J.B. (1959). Observations on the nutrition of the land planarian Orthodermus terrestris (O.F. Muller). Biol. Bull., 117, 119-124.

Jennings, J.B. (1962). Further studies on feeding and digestion in triclad turbellaria. Biol. Bull., 123, 571-581.

Jennings, J.B. (1968). Platyhelminthes, Nutrition. In M. Florkin and B.T. Scheer (Eds.), Chemical Zoology, Vol. 2. Academic Press, New York and London. pp. 640.

Jones, G.W., and I.D. Bowen (1979). The fine structural localization of acid phosphatase in molluscan pore cells. Suppl. Proc. Roy. Microsc. Soc., 14, A6-A7.

Jones, G.W., and I.D. Bowen (1980). The fine structural localization of acid phosphatase in pore cells of embryonic and newly hatched Deroceras reticulatum (Pulmonata : Stylommatophora). Cell Tiss. Res. (In press).

Kelley, E.G. (1931). The intracellular digestion of thymus nucleoprotein in triclad flatworms. Physiol. Zool., 4, 515-541.

Kerr, J.F.R., A.E. Wyllie, and A.R. Currie (1973). Apoptosis: A basic biological phenomenon with wide ranging implications in tissue kinetics. Brit. J. Cancer, 26, 239-257.

Kerr, J.F.R., B. Harmon, and J. Searle (1974). Cell deletion in tadpole tail. J. Cell Sci., 14, 571-585.

Metchnikoff, E. (1892). Pathologie Comparee de l'Inflammation. Masson, Paris.

Metchnikoff, E. (1901). L'Immunite dans les maladies infectienses. Paris (English translation by F.G. Binnie, 1905, Immunity in infective diseases. Cambridge University Press).

Novikoff, A.B. (1976). The Endoplasmic Reticulum: A cytochemist's view. (A review). Proc. Natl. Acad. Sci. U.S.A., 73, 2781-2787.

Osborne, P.J., and A.T. Miller (1962). Uptake and intracellular digestion of proteins (peroxidase) in planarians. Biol. Bull., 123, 589-596.

Osborne, P.J., and A.T. Miller (1963). Acid and alkaline phosphatase changes associated with feeding, starvation and regeneration in planarians. Biol. Bull., 124, 285-292.

Padawer, J. (1969). Uptake of colloidal thorium dioxide by mast cells. J. Cell Biol., 40, 747-760.

Pedersen, K.J. (1961). Some observations on the fine structure of planarian protonephridia and gastrodermal phagocytes. Z. Zellforsch., 53, 609-628.

Rose, C., and S. Shostak (1968). The transformation of gastrodermal cells to neoblasts in regenerating Phagocata gracilis (Leidy). Exp. Cell Res., 50, 553-561.

Rosenbaum, R.M., and C.I. Rolon (1960). Intracellular digestion and hydrolytic enzymes in the phagocytes of planarians. Biol. Bull., 118, 315-323.

Ryder, T.A., and I.D. Bowen (1974). The fine structural localization of alkaline phosphatase in Polycelis tenuis Iijima. Protoplasma, 79, 19-29.

Ryder, T.A., and I.D. Bowen (1975). The fine structural localization of acid phosphatase activity in Polycelis tenuis Iijima. Protoplasma, 83, 79-90.

Ryder, T.A., and I.D. Bowen (1977a). Endocytosis and aspects of autophagy in the foot epithelium of the slug Agriolimax reticulatus (Muller). Cell Tiss. Res., 181, 129-142.

Ryder, T.A., and I.D. Bowen (1977b). Studies on transmembrane and paracellular phenomena in the foot of the slug Agriolimax reticulatus (Muller). Cell Tiss. Res., 183, 143-152.

Saint-Hilaire, C. (1910). Beobachtungen uber die intracellulare Verdauung in den Darmzellen der Planarien. Z. allg. Physiol., 11, 177-248.

Straus, W. (1959). Rapid cytochemical identification of phagosomes in various tissues of the rat and their differentiation from mitochondria by the peroxidase method. J. Biophys. Biochem. Cytol., 5, 193-204.

Straus, W. (1962). Cytochemical investigation of phagosomes and related structures in cryostat sections of the kidney and liver of rats after intravenous administration of horse-radish peroxidase. Exptl. Cell Res., 27, 80-94.

Straus, W. (1967a). Changes in intracellular location of small phagosomes (micropinocytic vesicles) in kidney cells and liver cells in relation to time after injection and dose of horse-radish peroxidase. J. Histochem. Cytochem., 15, 281-393.

Straus, W. (1967b). Lysosomes, phagosomes and related particles. In D.B. Roodyn (Ed.), Enzyme Cytology. Academic Press, London and New York. pp. 239-319.

Willier, B.H., L.H. Hyman, and S.A. Rifenburgh (1925). A histochemical study of intracellular digestion in triclad flatworms. J. Morph. and Physiol., 40, 299-340.

FEEDING IN FRESHWATER TRICLADS—ADAPTATIONAL ASPECTS

P. CALOW

Department of Zoology, University of Glasgow, Glasgow, U.K.

ABSTRACT

All aspects of food-getting in freshwater triclads - including selection, searching, capture and ingestion - are reviewed. About each phase of the feeding process I ask: How does it take place? - and - What is its adaptive significance? Reference is made, on the one hand, to physiological and behavioural information and, on the other, to the evolutionary theory of optimal feeding strategies.

KEYWORDS

Optimality principles; energy maximization; time minimization; numbers maximization; searching strategies; gut-filling strategy.

INTRODUCTION

Freshwater triclads are suctorial carnivores which use a protrusible pharynx to feed from a variety of live and sometimes dead prey. They are liberal in their choice of foods but different species usually have specialities in the total spectrum of available prey (Reynoldson and Davies, 1970). The feeding process in these, as with many animals, consists of a series of meals (say of size M) eaten over a prescribed time (tm) and separated by intervals (ti) in which triclads search for more food or engage in other activities like copulation and laying eggs. This general scheme is illustrated in Fig. 1. In what follows I will consider how, in triclads, the selection, location and acquisition of food, as specified by the parameters M (amount eaten), tm (mealtime) and ti (time between meals), influence fitness. Or, to put it another way: Why have the feeding strategies that are now observed in freshwater triclads been favoured by natural selection?

Fitness is a complex term which expresses the number of viable descendents (ideally individual genes) that derive from organisms with particular genetically determined traits. It therefore depends on how these traits influence generation

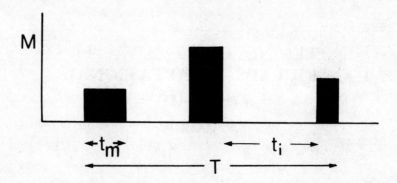

Fig. 1.   Schedule of food intake in discontinuous
feeders.

time, survivorship and fecundity.   Long term demographic effects, however, are
often difficult to quantify for minute by minute physiological and behavioural processes like feeding and more immediate measures of fitness are required.   One
which has been used extensively in the development of optimal foraging theory is
the net resource (usually, and as used below, resource = energy) returns from either
one or a series of meals (Krebs, 1978).   This is simply the difference between the
energy intake from a meal and the cost of obtaining and processing it (Schoener,
1971).   The gross intake (I) over a convenient length of time (T, Fig. 1) may also be
used but this measure assumes that the costs are either negligible or are constant
between strategies.   The assumption behind both indices is that fitness should increase with intake (energy maximization strategy; Schoener, 1971) because, in prin-
ciple, this should lead to increased growth rates, reduced generation times and
increased reproductive outputs.   On the other hand, some animals may have fixed
energy requirements (i. e. extra energy does not improve either developmental or reproductive performance) and should therefore be selected to reduce the time they
spend feeding (time minimization strategy; Schoener, 1971) rather than to maximize
their energy input.   However, since both the growth and reproductive performance
of triclads is sensitive to food supply (Calow and Woollhead, 1977a) it will be
assumed that they conform to the energy maximization principles and these will be
used, in what follows, to evaluate the adaptive value of feeding strategies.

In successive sections I consider which foods are and should be eaten, how they
are found and captured, and how ultimately they are ingested.   Each section reviews information from the literature and also introduces original data.   The
chapter has benefited from numerous discussions with research associates and
students and I am particularly grateful to Mr A. F. Davidson for contributing data
and ideas to the final section on the ingestion of food by triclads.

THE QUALITY OF FOODS EATEN

Interspecific Variation

In natural populations, triclads feed from a wide variety of prey (Hyman, 1951) and
this is expected in animals which exploit a poor food supply (MacArthur, 1972).
Despite the overall emphasis on generalization, however, Reynoldson and co -
workers have conclusively shown that in the field, British lake-dwelling triclads
do show some consistent variations in the foods eaten (see references in Table 1).
It is supposed that a particular species is competitively superior on the "favoured"
prey types and that these provide a refuge (Reynoldson, 1966) against competitors.
Laboratory experiments, that have monitored the survival of triclads on different
combinations of prey in the presence of a variety of competitors, have tended to
confirm this food refuge hypothesis (Reynoldson and Bellamy, 1971, 1973).   But
to what extent can refuges be considered as products of selection and do they conform to the energy maximization model?   In this context, the latter predicts that
species should be adapted to exploit their preferred foods more economically than
any others.

That triclads are competitively superior on their food refuges lends support to this
hypothesis but there remains the possibility that some triclads find refuge on foods
for which they are not optimally adapted because they are excluded from preferred
foods by even more efficient competitors.   In other words, biases observed in the
diets of triclads in the field may be a matter of expediency and have little to do
with optimal adaptation.   Indeed this kind of phenomenon is likely to be particularly important in the early stages of niche separation before the process of adaptation has caught up with shifts in diet.

Table 1 lists the food refuges for several, British lake-dwelling triclads.   Active
arthropods are taken, as prey, only by the large dendrocoelid species which have
a complex musculature and a well-developed anterior sucker for striking and holding prey (Hyman, 1951).   The much smaller planariids and the slightly smaller
dugesiids have no anterior sucker (Hyman, 1951) and concentrate on damaged and
freshly dead arthropods (Reynoldson and Davies, 1970) and on more sluggish
gastropods and oligochaetes.   These distinctions in food choice, which have also
been formulated more rigorously by Adams (1979) using ordination procedures, are
therefore related to morphological distinctions;   namely to the occurrence of an
anterior sucker and possibly also to the size of the triclads.   In the laboratory the
planariids and dugesiids grow and survive poorly on healthy Asellus, an active
isopod (Reynoldson and Young, 1963), but will grow and survive well on wounded
and immobilised Asellus (Calow and Woollhead, 1977a; Woollhead, 1979) and may
even be competitively superior to Dendrocoelum lacteum on this food (Reynoldson
and Bellamy, 1973).   Similarly, in the laboratory D. lacteum will not grow on
oligochaetes and it ignores gastropods (Reynoldson, 1975).

Such differences between the diet of the dendrocoelids and the other triclads are
clearly dependent on genetically determined differences in morphology which have
consequences for the economics of feeding.   Thus the metabolic cost of capturing
and restraining active prey without an anterior sucker is likely to make arthropods
unprofitable for planariids and dugesiids.   Alternatively, the low energy content of

TABLE 1   Food Refuges of British Triclads

Species	Presence of anterior sucker[+]	Size[**] (mm)	Food refuge	Activity of food[*]	Source
Planaria torva (Müller)	No	8 – 12	gastropods	+	Reynoldson and Sefton 1976 Reynoldson and Piearce 1979a
Polycelis nigra (Müller)	No	7. 5 –12	oligochaetes	0/+	Reynoldson 1978
Polycelis tenuis (Ijima)	No	8 – 12	"	0/+	Reynoldson and Bellamy 1974
Dugesia lugubris (Schmidt)	No	11 –17	gastropods	+	Reynoldson and Davies 1970, Reynoldson and Piearce 1979b
Dugesia polychroa (Schmidt)	No	"	"	+	
Dendrocoelum lacteum (Müller)	Yes	14 – 25	Asellus/other fast moving arthropods	+++	Bellamy and Reynoldson 1974
Bdellocephala punctata (Pallas)	Yes	35	same as Dendrocoelum lacteum plus molluscs	+++	Reynoldson 1978 Adams 1979

+   From Hyman 1951
*   0 = immobile;   + - +++ a subjective scale measuring increasing activity
**   From Reynoldson 1978 = length of gliding triclad

snails (they store glycogen rather than fat; Russell-Hunter and co-workers, 1968) together with their protective shell, and the small body sizes of oligochaetes (which package a small amount of biomass per individual) make these less profitable than the arthropods for the dendrocoelids.

Plasticity in food selection, relative to the presence of other species, is a particularly obvious feature of the detailed aspects of feeding biases in the dugesiids and planariids.   This is illustrated by the following observations: (1) P. tenuis and P. nigra eat significantly more snails when D. polychroa is absent (Reynoldson and Piearce, 1979b);   (2) D. polychroa and D. lugubris eat significantly more snails when P. torva is absent (Reynoldson and Piearce, 1979a);   (3) in the absence of Polycelis species D. polychroa thrives on oligochaetes (Boddington and Mettrick, 1974).   These modifications in diet are undoubtedly attributable to the presence and and possibly the action of competitors but it is not possible to determine from this kind of evidence if the shifts observed are the result of active decision processes on the part of the triclads or simply due to changes in availability of the foods

brought about by the action of competition.   In any event the shifts are likely to
be economically sensible for if the presence of a competitor so reduces the
availability and hence the rate at which a normally preferred food is encountered
then it may become less profitable (supplies less energy per unit time) than more
abundant but less rich (because it is less easily handled) alternatives (MacArthur,
1972).

Intraspecific Variation

Do triclads simply eat what they can relative to physiological and morphological
constraints and to food availability or can they exercise more precise discrimination in terms of their own internal states and external conditions?   Do they evaluate each potential item of prey they encounter on the basis of prior feeding experience or do they simply eat all the edible items they find?   Griffiths (1975) refers
to the latter as a numbers maximization strategy and suggests that because of their
limited neural capacity most invertebrates are likely to conform to this rather than
to the energy maximization principles.   However, Elner and Hughes (1978) have
shown that the selection of some foods by the shore crab Carcinus maenas conforms to what would be expected for energy maximization, and Vadas (1977) has
found that sea urchins seek out foods which maximize their growth rates and reproductive output and therefore their net energy input.   In this section I investigate whether triclads are capable of active discrimination using experiments on a
single species, D. polychroa, offered a choice between three size categories of a
single species of prey, Asellus aquaticus

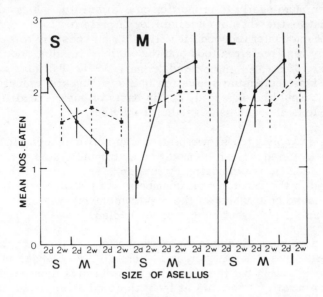

Fig. 2   Food selection by D. polychroa.   Solid lines are for 2-day
         starved triclads and broken lines for 2-week starved triclads.

Figure 2 shows the results of the choice experiments which offered equal quantities of the three categories of prey to three size classes of triclad.   The size classes of triclads (measured as plan area) were $S<20mm^2$, $M = 20-40mm^2$, $L>40mm^2$ and each was made up of two separate groups in different trophic states; one group having been starved for two days before the test and the other for two weeks.   The size classes of Asellus were: s<5mm, m = 5-10mm, l>10mm long.   A single tri-clad of appropriate size was offered a choice between the size classes of Asellus (4 individuals per class) arranged randomly around the edge of a glass trough (dia. = 8cm) containing 100 ml of filtered lake water.   This experimental regime was re-plicated five times for each trophic category of triclad within each size class.   All observations were carried out at 10^oC over a 14-hour period (0800hr to 2200hr) with eaten Asellus being noted and replaced at one hourly intervals.   The results are presented as the mean numbers of Asellus eaten for each triclad size class.   The 95% confidence limits are shown as vertical bars.

Despite the equal quantities and availabilities of the different prey categories the two-day starved triclads fed from some size classes more than others and the pre-ferences shifted with the size of the triclads.   Small Asellus were preferred by small triclads but medium and large triclads avoided this category and tended to prefer both medium and large Asellus.   Our observations suggest that this is roughly correlated with the growth-promoting powers of the prey in that small tri-clads grow best on small Asellus whereas medium and large triclads grow best on medium and large Asellus but grow only slightly on small Asellus.   The data do suggest, therefore, that D. polychroa can exercise some form of positive discrimi-nation on the basis of the maximization of net energy returns and hence growth rates.   Furthermore, this interpretation is reinforced and extended by the data on the two-week starved triclads.   Here preferences were much less obvious since prey types were eaten more nearly in proportion to availability.   It is to be ex-pected, however, that as the absolute availability of preferred food becomes re-duced and the rate of encounter reduced, it will become necessary to supplement the diet with normally less preferred food and the width of the diet should in-crease (Emlen, 1966, 1968; Eastabrook and Dunham, 1976; Pulliam, 1974; Schoener, 1971).   Since increasing starvation is a symptom of a reduction in the availability of food it would be expected, on the basis of energy maximization principles, to elicit less choosiness in feeding behaviour.

In conclusion, since D. polychroa shows preferences between such similar items of prey and since these preferences can be modified by trophic status there is strong evidence for some kind of active choosing by this species.   That is to say, the first prey encountered in the experimental chambers was not always eaten.   Whether it was, seemed to depend on the size of the prey (which was related in turn to its yield of net energy) and the state of hunger of the triclad.

DeSilva (1976a) has also published data on the selection of different size classes of Asellus by a triclad; this time D. lacteum.   Here all size classes of triclad seemed to prefer small Asellus but the larger triclads did take proportionately more large Asellus than the small.   There are at least three possible explanations for this result.   First, no information is given on the nutritive status of the triclads so they might have been very "hungry".   Second, the Asellus were active and the small ones might have been easier to capture than the large (i.e cost less in time and energy to capture as is, in fact, suggested by observation) so they would have been a more profitable food than the other size categories. Third, and finally, $\underline{D}$. lacteum is more of a sit and wait feeder than $\underline{D}$. polychroa (see below) and animals adopting this strategy are more likely to be numbers than energy maximizers (Griffiths, 1975).

TABLE 2   Effect of Increasing Density of Small (s) Asellus
           Relative to Medium (m) on the Food Choice of
           Medium Sized D. polychroa at 10°C

Relative abundance of Asellus	4m/4s	4m/8s	4m/16s
No. of medium eaten	2.0 ± 0.44	1.4 ± 0.26	1.0 ± 0.00
No. of small eaten	0.8 ± 0.18	1.2 ± 0.20	1.7 ± 0.18

Table 2 shows the influence of abundance of less preferred food relative to that of most preferred food on selection by medium-sized $\underline{D}$. polychroa offered a choice between small and medium Asellus. In each regime the absolute abundance of medium Asellus was kept constant at four per triclad but the number of small Asellus per chamber was varied. An analysis of variance on the data has shown that relative abundance of the two prey types did have a significant effect on the amounts of small and medium Asellus eaten (F $\backsim$ 3.0 for 2/12 d.f., P<0.05). The trends, as judged by confidence limits (P<0.05), were for small Asellus to be eaten to a greater extent and medium to a lesser extent as the abundance of the small Asellus was increased relative to the medium.

This result seems to argue against the previous conclusion since the evidence suggests that triclads were influenced more by encounters than by considerations of profitability.If predators recognise prey instantly and for each there is a constant handling time independent of the rate of encounter, then as long as the most profitable foods are sufficient to meet the metabolic demands of the feeder they should be selected even if the more profitable foods are more abundant (Krebs, 1978). However, many invertebrates, and this includes triclads, locate prey by chemosensory, mechanical and tactile cues, and these non-visual methods involve finite recognition times. Here, as the abundance of less preferred foods is increased relative to that of normally best preferred foods, so much time may be lost in checking the former that the profitability of the latter becomes seriously impaired. It may then become more profitable to feed preferentially from the usually inferior foods and this has been considered rigorously by Hughes (1979). The main conclusion, however, is that the effect noted in Table 2 is still consistent with the energy maximization principles but this now becomes less distinct from what would be expected on the basis of numbers maximization (Hughes, 1979).

If triclads are able to energy-maximize it must be possible to show that they have the neurological and sensory wherewithal to do it. Assuming that energy returns are associated with prey size, the minimum requirements would be ability (a) to judge the size of encountered food; (b) to "remember" previous nutritional history (which may simply be a matter of sensing internal hunger states); (c) to be able to control the feeding response on the basis of these two inputs.   These criteria do not appear to be beyond the physiological capabilities of triclads.

                    SEARCHING FOR FOOD

Between meals, when food is not immediately available, the most obvious response is to search for more.   However, since searching involves active movement it also involves an energy cost and for the optimal response this must be balanced against expected returns from finding food.   Animals may therefore choose between a sit and wait strategy (Type 1 predator of Schoener, 1971) and a search-out strategy (Type 2 predator of Schoener, 1971); these, of course, being extremes on what is probably a continuous range of options.   The choice will depend on both the proba-bility of finding food for a given effort (which depends on the density of food and its mobility) and on the metabolic properties of the feeders (Calow, 1977a).   In general those animals which feed on active foods are more likely to sit and wait than those animals which feed on sessile food.   Thus, because triclads incur heavy costs from mucous losses as well as elevated metabolism during active move-ment (Calow and Woollhead, 1977b) and because they can withstand long periods without food (Calow, 1977b) usually recovering quickly, completely and often with some compensation after feeding is restarted (Calow and Woollhead, 1977a), they would be expected to adopt a feeding strategy biased towards waiting rather than searching (Reynoldson, pers comm.).   However, within this broad category some triclads are more active searchers than others and this is related to the properties of the foods exploited by different species.

Observations on the patterns of activity and metabolism of triclads during starvation (Calow and Woollhead, 1977b) have suggested, for example, that D. lacteum, feeding on active prey, adopts a sit and wait strategy but that P. tenuis feeding on less active foods adopts a search-out strategy.   D lacteum is less likely to move and moves at a reduced rate as it becomes increasingly starved whereas P. tenuis responds to starvation in the opposite way by becoming more active.   More recent, as yet unpublished work (Calow, Davidson and Woollhead, in prep.), has suggested that this basic searching response is, in fact, more complex than was initially en-visaged.   Firstly, after 10 days starvation at $10^{\circ}C$ P. tenuis begins to become less active and after 14 days is very quiescent.   It can quickly be reactivated, however, by the presence of a wounded Asellus or even an extract from the Asellus, indica-ting that the reduction in activity is not attributable to metabolic exhaustion.   It is possible that having not been successful in locating prey over the first two weeks of starvation by the active searching strategy, P. tenuis switches to an energy conserving, sit and wait strategy and thereby conforms to the general pattern ex-pected for triclads (above).   Secondly, over the first week of starvation at $10^{\circ}C$ D. lacteum is poorly responsive to the reintroduction of Asellus or Asellus-extract in that the probability of triclads moving upon presentation of these stimuli is low (not zero).   This suggests that individuals are physiologically unresponsive to food and are perhaps in a "digestive pause" (Holling, 1963).   However, those D. lacteum which are activated in this period move rapidly and quickly capture prey and this is true of the majority of individuals after the first week.   It is likely, therefore, that there is both a physiological and behavioural component to the sit and wait response.

Results collected to date on other species suggest that B. punctata, another feeder on active prey (Table 1), adopts a similar strategy to D. lacteum, whereas P. nigra and the dugesiids, all feeders on less active prey (Table 1), adopt strategies more closely approximating to P. tenuis.

CAPTURE

When stimulated by the close proximity of food, and this can occur by both chemical and mechanical means (Bellamy and Reynoldson, 1974), triclads, if in the appropriate physiological state (see above), become more active.   Originally rather aimless movement often becomes directed as they sense chemical exudates escaping from the potential food.   Abnormal movements of wounded prey may also be important in attracting Polycelis species (Reynoldson and Young, 1963; Bellamy and Reynoldson, 1974).

When they make contact, the dendrocoelids quickly take hold of the prey using the anterior sucker, enfold them with their whole bodies and insert the pharynx.   In the planariids and dugesiids there is no anterior sucker and, as already discussed, this restricts the diet to less mobile prey.   According to Hyman (1951) some triclads use penis stylets to damage prey.

The testing of the suitability of food after discovery is vested mainly in the anterior margin of the triclad.   After extirpation of this margin, individuals find food as readily as before but fail to test or grip it and often glide away without ingestion (Hyman, 1951).   Brain removal inhibits ingestion in Crenobia alpina (Koehler, 1932). The tip of the pharynx may also be able to test the quality of the material extracted from the prey (Hyman, 1951).

Capture success may depend on external as well as internal factors   De Silva (1976a), for example, has shown that the capture success of Asellus by D. lacteum depends on cover.   Gravel of intermediate size provides better protection for Asellus than sand or large stones.

Mucous secretions also play a part in the capture of active prey by D. lacteum (De Silva, 1976b), and the aggregations of triclads, commonly encountered in nature, may enable the production of large mucous "traps" (Reynoldson and Young, 1963).   Mobile Asellus are certainly impeded when they make contact with a mucous patch (Jennings, 1957; per observation) and this is particularly so if small particles of sand are also present.   The latter become attached to the limbs of Asellus by the mucus (De Silva, 1976b).

There are two important, but as yet unanswered, questions about the involvement of mucus in the capture of prey.   First, given that the secretion of mucus is expensive (Calow, 1979; Calow and Woollhead, 1977b), do aggregations of triclads favour "cheats" who use the mucus of others rather than their own?   This might be favoured by selection but it would not necessarily be an evolutionarily stable strategy (Calow, 1979).   Second, do the search-out feeders like P. tenuis, which exploit immobile prey, make use of mucous secretion and aggregation to a lesser extent than sit and wait triclads?   This would be expected since "trapping" plays a less important role in these species than active searching.   T.B. Reynoldson thinks mucus is less important for prey capture by the Polycelis species (pers. comm.).

AMOUNT EATEN (M)

The maximization of both gross and net returns from food requires the maximization of m/tm (significant increases in metabolic costs as a result of increasing feeding rates being unlikely).   Optimal foraging theory would predict, therefore, that once a meal (food patch) has been located feeding should be at a high and constant rate (Krebs, 1978).

The food extraction curve for D. polychroa on individual Asellus is presented in Fig. 3.   Data were obtained by allowing triclads to feed from pre-weighed (wet)

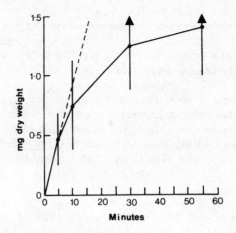

Fig. 3.   Dry weight extracted from large Asellus by large
           D. polychroa after varying tm at 10°C.   Broken
           line is theoretical expectation.

Asellus at 10°C.   Triclads were removed after differing tm and the Asellus was dried.   An initial dry weight was predicted for the Asellus from DRY WT = 0.03 + 0.28 WET WT (an equation derived from over 100 observations).   The difference in dry weight before and after feeding was taken to be that consumed though some material would have leaked into the water.   Ten replicates were used per tm and vertical bars represent 95% confidence limits.   The broken line is the theoretically expected relationship and, as is clear, the actual results rather differed from it. Ingestion rate decreased continuously with time.

Figure 4 shows data from experiments in which individual triclads were presented with a succession of fresh Asellus at 10-minute intervals over a continuous feeding period.   Many triclads would not refeed after transference and so there are only results for 3 individuals (marked 1 to 3) out of more than 50 tested.   Nevertheless, the data do indicate that no decline occurred in ingestion rate under these

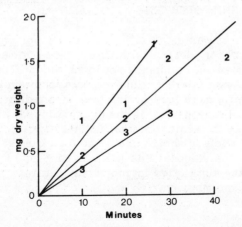

Fig. 4.   As in Fig. 3 but using a different technique
               (see text).

circumstances and suggest that the deviation from prediction in Fig. 3 was a "food"
and not a "feeder" effect.   Some property of the Asellus made the last part of the
meal more difficult to extract than the first and this sort of effect has been noted
for many animals, including other suctorial feeders (Cook and Cockrell, 1978).

The amount eaten per meal by triclads is also proportional to body size.   The few
data available on this relationship suggest a significant ($P<0.05$) linear relationship between triclad plan area (and hence surface area; Woollhead, 1979) and food
input (Fig. 5).

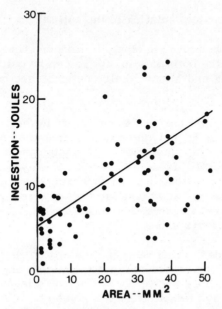

Fig. 5. Joules (= mg dry weight x 20) eaten by $\underline{D}$. lacteum at
               $10^{\circ}C$.   Data supplied by A. S. Woollhead.

RELATIONSHIP BETWEEN M, tm AND ti

In the previous section it was established that as $\underline{D}$. $\underline{polychroa}$ extracts food from its prey the returns probably decline with time. The optimal time to spend feeding at each meal to maximize I/T and hence net input can be found graphically (Krebs, 1978) as in Fig. 6. Assuming that triclads attack many prey over interval T and spend all ti searching for more food (which is an approximation - see above) then the average food intake for a habitat is the average intake per prey ($\hat{M}$) divided by the sum of the average inter-catch interval ($\hat{ti}$) and average meal time ($\hat{tm}$). In order to maximize I/T triclads should choose to feed from each prey long enough to maximize the slope of the diagonal line (i.e. $\hat{M}/(\hat{ti} + \hat{tm})$) in Fig. 6 and this occurs when it just touches the cumulative intake curve. At this point an optimal feeder should cease feeding and move on.

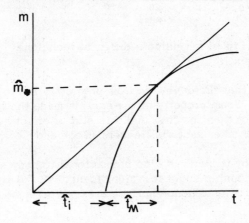

Fig. 6. A graphical solution to the optimal tm

If all that varies from one habitat to another, or from one time to another is food availability and hence $\hat{ti}$, the optimal value for $\hat{tm}$ can be calculated from Fig. 6. Clearly as $\hat{ti}$ increases $\hat{tm}$ and $\hat{M}$ should also be increased but not in a linear fashion.

Figure 7 shows the relationship between ti, tm and M in three species of freshwater triclads. Ingestion in $\underline{D}$. $\underline{polychroa}$ was measured as for the results in Fig. 3 and is in mg dry weight. Ingestion in $\underline{D}$. $\underline{lacteum}$ and $\underline{P}$. $\underline{tenuis}$ was assessed using a scoring technique (1. feeding observed but only small amount of body contents removed; 2. all body contents removed except the gut; 3 all body contents removed including the gut) and these latter data were kindly supplied by Mr A. Davidson. Vertical bars represent confidence limits. It was not possible to put limits on the non-parametric scores.

As predicted tm and M tended to increase, particularly in $\underline{D}$. $\underline{polychroa}$, as ti increased. (These data will be considered in detail elsewhere - Calow, Davidson, Woollhead, in prep.). Another mechanism which would lead to the same effect, however, is if feeders always fed or attempted to feed to satiation at each meal; i.e. the "gut-filling model" of Holling (1963). This suggests that at each meal feeders should feed to fill their guts so that as ti is increased the space in the gut

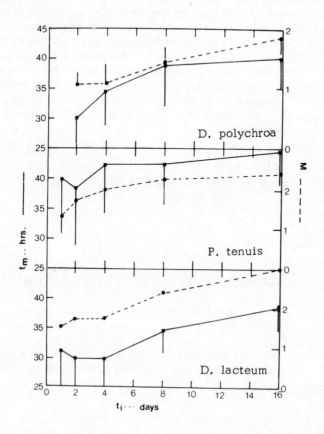

Fig. 7.   Time spent on prey (tm) and the size of a meal (M)
          for 3 species of triclad after different starvation
          times (ti).

is increased and the amount required for satiation (M) is increased.   Hence M and
tm should increase in a non-linear fashion (Holling, 1963) with ti.

It is important to note that a "gut-filler" need not be an "optimizer" since those
animals which waste time feeding to satiation may not gain so much, in the long
term, as those which, though never filling their guts, feed frequently on easily
accessible material.   However, gut-filling will approximate to optimization as $\hat{t}i$
is increased and the amount of food available at each meal is equal to or greater
than the amount required to fill the gut.

As Cook and Cockrell (1978) have pointed out, the crucial difference between optimizers and gut-fillers is that the former should respond to average inter-catch
intervals (i. e. $\hat{t}i$) and the latter to the inter-catch interval immediately preceding

a particular meal (i.e. ti) because it is the latter which determines gut capacity. Thus, if ti differs markedly from t̂i (based on the average experience of the feeder in a particular habitat) the gut-filler should feed in proportion to ti and the optimizer to t̂i.   Only work on D. polychroa has been carried out to distinguish between these two possibilities.   Groups of ten triclads (medium sized) were fed either once every 2 (1/2) or 6 (1/6) days on Asellus (medium sized) at $10°C$. Another group was fed on a randomly altering 2- or 6-day pulse under exactly the same conditions.   The mean amounts of Asellus eaten per meal over a one-month period (approximately two months after the experiment had been started) are given in Table 3.   Analysis of variance showed that there were very significant differences between the means.   As judged by the overlap of confidence limits there were

TABLE 3   Mean Dry Weight of Asellus Extracted by
           D. polychroa on Different Pulse-Feeding
           Regimes

Regime	1/2	1/6	1/2 or 6	
			1/2	1/6
	$0.66±0.14$	$1.48±0.48$	$0.47±0.10$	$1.66±0.55$
	$F = 6.10$ for 3/81 d.f.		$P<0.05$	

no significant differences between the tm's and M's for the same values of ti irrespective of whether triclads had been fed on the continuous or randomly altering regimes.

After this initial series of observations triclads on the continuous regimes were transferred from a 2- to a 6-day pulse and vice versa, and the next M was measured and is given in Table 4.   There were significant differences in these data and confidence limits suggest that the values of M after transference did not differ significantly from the M's of worms which had been kept continuously on a 2- or 6-day pulse.

TABLE 4   Mean Dry Weight of Asellus Extracted by
           D. polychroa After Transference from One
           Pulse-Feeding Regime to Another.

Regime	1/2	1/6	1/2 to 1/6	1/6 to 1/2
	$0.66±0.14$	$1.48±0.48$	$1.81±0.62$	$0.73±0.19$
	$F = 4.2$ for 3/63 d f.		$P<0.05$	

These results suggest that D. polychroa was acting as a gut-filler since M was
more responsive to ti than t̂i.   That is to say, the triclads on the alternating
regime, if optimizers, would have been expected to feed throughout on the basis of
a t̂i of between 2 and 6 rather than a ti equal to either 2 or 6.   Furthermore, feeding levels of the 1/2 and 1/6 groups on the alternating series should not have
differed significantly.   In the event they did.   Similarly in the transference experiments optimizers should have responded to long-term experience rather than on the
basis of the immediate experience after transference.   Again the latter was found
to be more potent than the former

DISCUSSION

In the selection and acquisition of food, freshwater triclads conform, approximately,
to the energy maximization principles.   However, because of their "primitive"
searching method (which involves the use of tactile and chemosensory rather than
visual cues) and possibly because of their "primitive" nervous system (which may
not allow the storage and detailed integration of information needed to compute
optimal responses) their food choice is influenced by relative prey abundance
though, perhaps surprisingly, some adaptable decision processes seem to be involved, and they "gut-fill" rather than "optimize" as an ingestion strategy.   In
situations where ti is large, as it is likely to be in triclads which exploit a poor,
carnivorous food supply, gut-filling may approximate to an optimal response.
Constraints of biological organization therefore mean that the actual feeding behaviour of triclads differs in means, if not in end-result, from the one expected on the
basis of the energy maximization model.   Elner and Hughes (1978) have shown that
the relative abundance of food is also important in determining the diet of the shore
crab Carcinus maenas and this may be a usual feature of invertebrate feeding
strategies (Griffiths, 1975; Hughes, 1979).   Data presented by De Silva (1977a)
for D. lacteum (her Table 2) suggest that relative prey abundance is also important
in determining the diet of these sit and wait feeders.   There are few other data on
gut-filling versus optimization in the lower invertebrates but a gut-filling strategy
is likely to be associated with simple nervous systems.   However, an optimization
as opposed to a gut-filling strategy has been recorded at the arthropod level for
Notonecta, a suctorial predator, by Cook and Cockrell (1978).

Energy maximization principles are also useful in understanding differences in food
choice and acquisition strategies between species.   The large dendrocoelids, with
an anterior sucker, exploit active prey and are able, therefore, to adopt a sit and
wait strategy.   The smaller species feed on less active prey and even freshly dead
foods and, in consequence, adopt a more active searching strategy.   This distinction may have consequences for the metabolism of the different species and ultimately for their life-cycles (Calow, Davidson and Woollhead, in prep).   The
search-out feeders undoubtedly invest more energy per unit of food ingested in
metabolism than the sit and wait feeders and this means they are less efficient con-
verters of food to somatic and gametic tissue than the dendrocoelids.   As a result
they tend to adopt a repeated breeding strategy (i.e. are iteroparous) whereas
the dendrocoelids tend to breed once only, putting a lot of effort into reproduction,
and die thereafter (i e. are semelparous).   Thus it can be seen that the adaptational
aspects of feeding behaviour may have consequences for the organism as a whole
and, in particular, for the adaptational aspects of life-cycle strategies.

REFERENCES

Adams, J. (1979). The population ecology of Bdellocephala punctata (Pallas), a freshwater triclad. Unpubl. Ph. D. thesis, University of Liverpool.

Bellamy, L. S., and T. B. Reynoldson (1974). Behaviour in competition for food amongst lake-dwelling triclads. Oikos, 25, 356-364.

Boddington, M. J., and D. F. Mettrick (1974). The distribution, abundance, feeding habits and population biology of an immigrant triclad Dugesia polychroa (Platyhelminthes : Turbellaria) in Toronto Harbour, Canada. J. Anim Ecol., 43, 681-689.

Calow, P. (1977a). Evolution, ecology and energetics: a study in metabolic adaptation. Adv. Ecol. Res, 10, 1-61.

Calow, P. (1977b) The joint effect of temperature and starvation on the metabolism of triclads. Oikos, 29, 87-92.

Calow, P. (1979). Why some metazoan mucus selections are more susceptible to microbial attack than others. Amer. Natur, 114, 149-152.

Calow, P., and A. S. Woollhead (1977a). The relationship between ration, reproductive effort and age-specific mortality in the evolution of life-history strategies; some observations on freshwater triclads. J. Anim Ecol., 46, 765-781.

Calow, P, and A. S. Woollhead (1977b). Locomotory strategies in freshwater triclads and their effects on the energetics of degrowth. Oecologia, 27, 353-362.

Cook, R. M., and B. J. Cockrell (1978). Predator ingestion rate and its bearing on feeding time and the theory of optimal diets. J. Anim. Ecol., 57, 529-548.

De Silva, P. K (1976a). The factors affecting the feeding of Dendrocoelum lacteum (Müller) (Turbellaria, Tricladida) on Asellus aquaticus (L.) (Crustacea, Isopoda). Arch. Hydrobiol., 77, 347-374.

De Silva, P. K. (1976b). The importance of mucus of Dendrocoelum lacteum (Müller) (Turbellaria, Tricladida) in community existence Arch. Hydrobiol., 78, 363-374.

Eastabrook, G. F., and A. E. Dunham (1976). Optimal diet as a function of absolute abundance, relative abundance and relative value of available prey. Amer. Natur., 110, 401-413.

Elner, R. W. and R. N. Hughes (1978). Energy maximization in the diet of the shore crab Carcinus maenas (L.) J. Anim. Ecol., 47, 103-116.

Emlen, J M. (1966). The role of time and energy in food preference. Amer. Natur. 100, 611-617.

Emlen, J. M. (1968). Optimal choice in animals Amer. Natur., 102, 285-390.

Griffiths, D (1975). Prey availability and the food of predators Ecology, 56, 1209-1214.

Holling, C. S. (1963). An experimental component analysis of population processes. Mem. Ent. Soc. Can., 32, 22-32.

Hughes, R. N. (1979). Optimal diets under the energy maximization premise: The effects of recognition time and learning. Amer Natur., 113, 209-221.

Hyman, L. H. (1951). The Invertebrates, Vol. II, McGraw-Hill Book Co., New York.

Jennings, J. B. (1957). Studies on feeding, digestion and food storage in free-living flatworms (Platyhelminthes: Turbellaria). Biol. Bull., 112, 63-80.

Koehler, O. (1932). Beitrage zur Sinnesphysiologie der Süsswasserplanarien.
    Z. Vergl. Physiol., 16, 606-756.
Krebs, J.R. (1978). Optimal foraging: decision rules for predators In J.R. Krebs
    and N.B Davies (Eds), Behavioural Ecology. Blackwell Scientific Publica-
    tions, Oxford   pp. 23-63.
MacArthur, R.H. (1972). Geographical Ecology. Harper and Row, New York, London.
Pulliam, H.R (1974). On the theory of optimal diets. Amer. Natur., 108, 58-74.
Reynoldson, T.B. (1966). The distribution and abundance of lake-dwelling triclads
    - towards a hypothesis. Adv. Ecol. Res., 3, 1-71.
Reynoldson, T.B. (1975). Food overlap of lake-dwelling triclads in the field.
    J. Anim. Ecol., 44, 245-250.
Reynoldson, T.B (1978)   A Key to the British Species of Freshwater Triclads.
    Freshwater Biological Association (U.K.) Sci. Publ. No. 23, 2nd (revised)
    edition.
Reynoldson, T.B., and L.S Bellamy (1971). Intra-specific competition in lake-
    dwelling triclads. A laboratory study. Oikos, 22, 315-328.
Reynoldson, T.B., and L.S Bellamy (1973). Inter-specific competition in lake-
    dwelling triclads. Oikos, 24, 301-314.
Reynoldson, T.B., and L.S. Bellamy (1974). Triclads (Turbellaria, Tricladida) as
    predators of lake-dwelling stonefly and mayfly nymphs. Freshwat. Biol.
    4, 305-312.
Reynoldson, T.B., and R.W. Davies (1970). Food niche and coexistence in lake-
    dwelling triclads. J. Anim. Ecol., 39, 599-617.
Reynoldson, T B , and B. Piearce (1979a). Predation on snails by three species of
    triclad and its bearing on the distribution of Planaria torva in Britain.
    J. Zool. in press.
Reynoldson, T.B., and B. Piearce (1979b). Feeding on gastropods by lake-dwelling
    Polycelis in the absence and presence of Dugesia polychroa (Turbellaria,
    Tricladida). Freshwat. Biol., 9, 357-367.
Reynoldson, T.B., and A.D. Sefton (1976). The food of Planaria torva (Müller)
    (Turbellaria-Tricladida). Oecologia, 10, 1-16.
Reynoldson, T.B., and J.O. Young (1963). The food of four species of lake-dwell-
    ing triclads. J. Anim. Ecol., 32, 175-191.
Russell-Hunter, W.D., Meadows, R.J , Apley, M.L , and Burky, A.J. (1968). On
    the use of a wet oxidation method for estimates of total organic carbon in
    mollusc growth studies. Proc. malac. Soc. Lond., 38, 1-11.
Schoener, T.W. (1971). Theory of feeding strategies. A. Rev. Ecol. Syst., 2,
    369-404.
Vadas, R.L. (1977). Preferential feeding: an optimization strategy in sea urchins.
    Ecol. Monogr., 47, 337-371.
Woollhead, A.S. (1979). Energy partitioning strategies in triclad flatworms with
    different life-cycles. Unpubl. Ph.D. thesis, University of Glasgow.

SELECTION IN SPONGE
FEEDING PROCESSES

T. M. FROST

*Department of EPO Biology, University of Colorado, Boulder,
Colorado 80309, U.S.A.

ABSTRACT

By considering the potential levels for variation in material processing by suspension feeding organisms, I develop a framework for considering selectivity
in their food handling. I use this framework to consider the feeding processes
of sponges. Using a species of freshwater sponge from a large tropical lake I
test several hypotheses generated from the selection framework.

Large scale selective processes occur in the sponge due to variation in clearance
activities. Selection at the individual particle level occurs due to differential
handling of materials by the digestive system. The parazoan nature of the sponge
feeding system permits the separate treatment of individual particles within
digestive cells. Digestive system transit time varies with food particle type.

KEYWORDS

Freshwater sponges; Lake Valencia; Porifera; Selective Feeding; Suspension
Feeding; Sponge Feeding; Sponges.

INTRODUCTION

Sponges, like other suspension feeders, are confronted with a milieu containing a
great variety of filterable materials. An understanding of the way that
organisms deal with this variety of substances is fundamental to a comprehension
of their feeding biology. In this report I outline a framework for considering
the variety of methods employed by all suspension feeders in handling filtered
matter. I consider demosponge feeding activities in regard to this framework
and test its applicability in a series of experiments utilizing a species of
freshwater sponge from a large tropical lake, Lago de Valencia, in Venezuela.

Substances filtered by suspension feeders vary in size and shape, in resistance
to digestion, and in nutritional value. Discrepancies exist between the
occurrence of materials in the habitats of organisms and their utilization by
these organisms. Here I define the processes causing these discrepancies as
selection including phenomena considered both mechanical and behavioral by
previous authors (Donaghay and Small, 1979).

Selective activities can be accomplished at a variety of levels in the feeding processes of suspension feeding organisms. Variation can occur:

1) In the generation of the stream of water to be filtered.
2) In pre-filtration screening processes.
3) In characteristics of the filtering apparatus.
4) In the handling of materials after initial filtration, but prior to digestive processes.
5) In digestive activities.

Considering each of these levels and summing their effects for the diversity of materials in a habitat defines the process of selection for an organism. I intend the listing above to be exhaustive, applicable to all suspension feeders, and some categories bear only minor relevance to sponges as will be seen below.

From this selection framework I have generated a basic hypothesis that there should be an overall balance, in organisms, of discriminating processes. If a suspension feeder selectively removes only a small portion of the spectrum of materials surrounding it, the incorporation of these substances should be relatively efficient and uniform. In contrast I would predict that an organism which takes up materials indiscriminantly would select from these substances by differential digestion. Viable passage of algae (Porter, 1975) represents an extreme of this case. Some organisms may combine a lower level of discrimination at a number of levels. By considering past research and the physiology of sponge feeding, I have generated several hypotheses regarding the levels and relative magnitudes of selective processes by the sponge.

At the first potential level for variation, a sponge could alter its rate of water transport in response to materials in its incurrent stream. Reiswig (1971a) directly measured water pumpage by a number of species of large reef sponges. He found a number of patterns in the variation of pumpage rates but these do not suggest a response to food availability. With smaller sponges, as with a great number of suspension feeders, it is not practical to determine water transport rates directly. Instead, water pumpage (1) and filtration activity (2 and 3) are grouped together and determined indirectly by a clearance rate based on material removal and an assumption of a filter efficiency value (see for e.g. Frost, 1976; 1978). In previous work with another species of freshwater sponge, Spongilla lacustris, I found that clearance rate was independent of particle type and concentration except under extreme conditions (Frost, in preparation). There appears to be little potential for variation at the actual filters of the sponge. Most of the flow of water must pass through the collars of the choanocytes which provide a mesh of about 0.1 μm in width (Rasmont and others, 1959; Fjerdingstad, 1961). I hypothesize that the uptake portions of the sponge feeding processes (1, 2, and 3) are relatively nonselective. I test this by examining the clearance activity of sponges on four particle types.

Once material has impacted on the filtering surface of a sponge it is phagocytized and transferred to amoebocytes which move throughout the mesohyle. Digestion occurs within these cells or after subsequent transfers to other similar cells (van Weel, 1949). An amoebocyte operates on a single or small number of particles. This process is a combination of categories 4 and 5 in the selection framework although I included 4 primarily to describe separate handling systems such as are exhibited by molluscs, polychaetes, and tunicates (Jørgensen 1966).

When one particle type is present at very high concentrations the parazoan digestive system of the sponge is capable of discriminating rapidly among materials. S. lacustris released yeast which were present at very high concentrations while continuing to filter and retain, at normal rates, bacterial cells present at near natural concentrations. Other evidence has suggested that this discrimination occurs under natural conditions. Reiswig (1971b) found very low retention of bacteria in marine sponges at certain times while normal uptake of other particles occurred. I have previously found evidence for a variation in turnover rates for different natural particles in the Valencia sponge (Frost, in preparation). I hypothesize that the bulk of selection by sponges occurs through differential handling of particles by the amoebocytes. I test this by examining the concentration of several natural particle types within the sponge and the change in these concentrations with time in sponges which have no input of new materials.

Ultimately, it is necessary to consider the actual incorporation of materials after digestive processes. The magnitude of selection at this level is difficult to quantify. Selection has clearly occurred when an organism removes and digests only one or a few of the particles from its feeding suspension. However, selective processes have also occurred when particles of greatly varied resistances either differentially pass through a feeding system or are incorporated at rates which do not reflect their relative resistances. Unfortunately, resistance in this sense is tautologically linked with digestion and is difficult to define in an absolute sense. Selection of this nature could best be examined by comparisons between organisms feeding on the same particle types but this has not been possible in my study at this time. I hypothesize that sponges should incorporate a variety of material at relatively uniform rates due to the differential handling of particles in the amoebocytes. I test this by measuring for four particle types, the amount of material filtered in one hour that is retained after 12 and 24 hours.

Combining the three hypotheses regarding sponges listed above I hope to generate an overall understanding of the sponge feeding system and to examine the utility of the selection framework I have proposed.

MATERIALS AND METHODS

I conducted this work with sponges from Lago de Valencia, Venezuela ($10^{o}10'$ N, $67^{o}45'$ W). Lewis and Weibezahn (1976) provide a detailed description of the lake. It is highly eutrophic with an abundant and varied phytoplankton population.

In extensive sampling I have observed only one species of sponge in the lake. It occurs throughout the lake primarily encrusting hard substrates but occasionally attached to aquatic macrophytes. The taxonomy of freshwater sponges is based on the structure of gemmule spicules (Penney and Racek, 1968). I have never observed gemmulation or vestigal gemmuloscleres in these sponges and a species identification is not currently possible. It exhibits characteristics of the genus Ephydatia.

To examine clearance rates and retention of materials I labelled a number of particles with radioisotopes. The particles included the yeast Rhodotorula glutinis (a sphere, approximately 4 μm in diameter with occasional budding) and 3 algae: Ankistrodesmus sp (a needle like cell, 80 μm long and 2-4 μm wide), Scenedesmus sp (a group of two or four rounded cells with spines extending from its ends, 45 μm long and 30 μm wide including spines), and Staurastrum sp (a pair of star shaped semicells, 30 μm long and 30 μm wide). I labelled the algae with ^{32}P following techniques described by Starkweather and Gilbert (1977). For R. glutinis I modified a medium for ^{32}P incorporation to facilitate ^{35}S uptake. It contained 0.2 g $CaCl_2$, 1.0 g NH_4Cl, 0.5 g KH_2PO_4, 2.25 g dextrose, 1.0 g bacto-peptone, and 1.0 g yeast extract per liter. Cultures were grown for two days at 27° C with shaking. I determined cell concentrations using a haemocytometer.

I gathered sponges attached to small stones from the shore of the lake and placed them at a convenient location in water approximately 1 m deep. I moved sponges to this location at least one day prior to an experiment. To determine clearance and retention rates I placed a number of sponges in glass aquaria with approximately 8 liters of lake water. The aquaria were placed at the lake shore in shallow water or immediately above the lake when wave action was high. After a one hour acclimatization period I added a suspension of labelled cells to provide the following final concentrations: R. glutinis, 1250 cells/ml; Ankistrodesmus sp, 1000 cells/ml; Scenedesmus sp, 200 cells/ml; and Staurastrum sp, 100 cells/ml. The differences in concentration are intended to reflect, approximately, the biomass of the different cell types. In most experiments one type of labelled cell was added to an aquarium. However for separate simultaneous determinations of clearance on two particles, they were mixed prior to introduction. The feeding suspension was stirred gently at 15 minute intervals for 1 hour and calibration samples of the suspension were taken at 15 and 45 minutes. After the feeding period the aquaria were flooded with lake water and sponges for clearance rate determinations were removed. I took 3 replicate samples of each of three or four sponges by slicing portions of them from the rocks they encrusted using a razor blade. Sponges for retention determinations were moved from the aquaria to enclosures of 1 cm width plastic mesh placed in the lake at a depth of 1 m. These sponges were then sampled as described above after 12 and 24 hours. In all experiments described above I never removed sponges from the water prior to sampling. All experiments were initiated between 8:00 and 11:00 hours during August-October, 1979 and water temperature was approximately 28° C throughout.

Sponge samples were placed in pre-weighed scintillation vials and then dried at 75° C for 24 hours. They were weighed and then digested in 1 ml of H_2O_2:$HClO_4$ solution (Bogden, 1976). Calibration measures for the feeding suspensions were obtained by filtering 1 ml from the aquarium suspension onto a 0.22 μm membrane filter. These filters were then treated as sponge samples except that drying time was 1 to 2 hours. I then added 10 mls scintillation fluor (808E, Riedel-de Haën AG, Seelze-Hannover) to all samples. They were then processed in a Packard Tri-Carb liquid scintillation counter using counting techniques suggested by Horrocks (1974).

Clearance rates were calculated by dividing the total counts in the sponges samples by the mean calibration values for the feeding suspensions and converting to a per dry weight basis. The sponges did not significantly affect the concentration of labelled cells in the aquaria during the experiments. For retention calculations I contrasted the total activity per dry weight after 12 and 24 hours with the mean activity accumulated by sponges in the 1 hour feeding period.

To examine the occurrence of particles within the sponge I utilized a method for "gut-content" analysis described in detail elsewhere (Frost, in preparation). Basically it involves macerating a pre-weighed portion of sponge very finely using a razor blade, adding water and agitating vigorously, finally preserving with Lugol's solution. Cells in this suspension were identified using information provided by W. M. Lewis (Lewis and Riehl, in preparation) and counted using a haemocytometer. I examined the occurrence of diatoms in a similar fashion, following techniques for spicule preparation (Penney and Racek, 1968) and counted diatom frustules among the spicules. For diatom analyses I considered only the relative abundance of forms and did not obtain actual concentrations. I examined the digestive activities on these particles by placing the sponge in filtered lake water (Whatman GF/C), allowing normal sponge activities but without the input of new materials. I conducted these experiments in a similar fashion to the isotope studies except that sponges were removed from the lake and briefly rinsed with filtered water prior to placement in the aquaria to remove any attached cells. The water in the aquaria was gently mixed at frequent inter-vals to prevent oxygen depletion.

To observe living sponges, I collected specimens in buckets without lifting them into the air. I examined specimens initially less than one hour after collection using a dissecting microscope at 25, 50 or 100 magnifications.

Aside from the geometric mean regression method (Ricker, 1973), I followed stati-stical techniques described in Sokal and Rohlf (1969). To compare clearance rates of sponges under different experimental conditions I used a two level nested analysis of variance. Other tests are described where applied.

RESULTS AND DISCUSSION

Clearance Activities

Fig. 1 illustrates the values I obtained for clearance rates in experiments with a single particle type. A substantial amount of variation occurs at each level in these experiments. Considering all of the data (coefficient of variation = 75.7%), the variation is partitioned as follows: within sponges, 19.6%, between sponges within experiments, 37.5%; and between experiments, 42.8%. Significant differences occur both between experiments (F = 3.92; df = 5,12; p < .025) and between sponges within an experiment (F = 6.75; df = 12, 36; p < .001).

Of the three separate yeast experiments, the second (Y2, middle group) was con-ducted in conditions most similar to the three algal experiments. I conducted it two days after the algal experiments which were all undertaken on the same day. Other than differences in particle concentration, protocols were the same. The first yeast experiment (Y1, left group) was conducted 44 days earlier than the algal experiment but otherwise in the same fashion. Data on retention for yeast were obtained at this time also. Clearance rates for Y1 are significantly greater than Y2 (F = 16.0; df = 1, 4; p < .025). The final yeast experiment (Y3, right group) was conducted in conjunction with Y2 and the diatom turnover experiment discussed below. I tested the effect of a lack of input to the sponge by measuring clearance rates after 4 hours in filtered lake water versus control sponges (Y2) which fed for one hour at the start of the 4 hour period. A large amount of within sponge variation in Y3 masks any difference between it and Y2. However the variation in Y3 and the great difference in the means for the two groups (Y2 = 425 and Y3 = 636 µl/sec/g dry wgt) suggest that the lack of input for 4 hours did affect the sponges.

Contrasting the Y2 values with the clearance rates for algae yielded different results than the overall analysis. As previously, there were significant differ-ences between sponges on one particle type (F = 4.36; df = 8, 24; p < .005) but between particles the variation was not statistically significant (F = 3.72; df = 3, 8; p < .10) although it still suggested differences.

The particle types in Fig. 1 are oriented in order of increasing size. The distribution of these data suggested an inverse relationship between particle size and clearance rate. I tested this using a Kendall rank correlation for the mean value for each sponge versus a ranking of particle size. For all of the data and for Y2 alone combined with the algal values the relationship was highly significant (tau = 1.06 and 1.78 respectively; p < .001 for both).

Overall, the data obtained separately for the variety of particle types clearly suggested differences in their rate of clearance, particularly for R. glutinis versus Staurastrum sp or Scenedesmus sp. To examine this situation I utilized

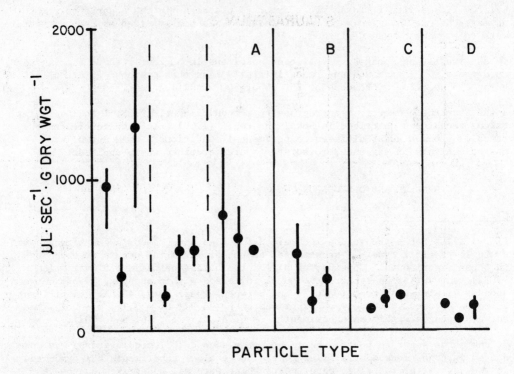

Fig. 1.   Clearance rates for sponges on; A. <u>Rhodotorula</u>
<u>glutinis</u>, B. <u>Ankistrodesmus</u> sp, C. <u>Staurastrum</u> sp,
and D. <u>Scenedesmus</u> sp.   Each point represents the mean of
3 subsamples of one sponge and the bars represent their range.
The lack of a bar indicates that the range was smaller than the
size of the symbol.   For <u>R</u>. <u>glutinis</u> three different
experiments are shown separated by broken lines (differences
are explained in the text).   Each set of three points represents
sponges from the same experimental conditions.

double-label isotope techniques to obtain clearance rates separately but simultaneously for the yeast and each of these algae.   Aside from providing a direct
test of differences in clearance rates, this technique permits separation of two
possible reasons for the differences.   The presence of the algae might slow down
the entire filtering process.   In this case the clearance of yeast would be
reduced concomitantly.   Alternatively, the feeding system could operate differentially on two particle types.

Figure 2 shows the results of the double-label analyses.   Both experiments were
conducted in the same 2 hour period.   For <u>Staurastrum</u> sp and <u>R</u>. <u>glutinis</u> there is
a tendency for slightly higher rates on yeast than on algae but this difference
is not significant $F = 1.56$; $df = 1,6$; $p < .50$).   Interestingly, the rates on
<u>Staurastrum</u> sp in this experiment are significantly greater than for the experi-
ment in Fig. 1 ($F = 6$, $df = 1,5$; $p < .05$).   For <u>Scenedesmus</u> sp and <u>R</u>. <u>glutinis</u>
differences in clearance are highly significant ($F = 19.2$; $df = 1,6$; $p < .005$).
A strong relationship exists between the simultaneously determined clearance
rates on yeast and either algal form.   Geometric mean (Model II) regressions
in both cases are highly significant (C.R. on yeast = 1.45 x C.R. on <u>Staurastrum</u>
- 0.2884; $t = 14.9$; $df = 11$; $p < .001$ and C.R. on yeast = 2.61 x C.R. on

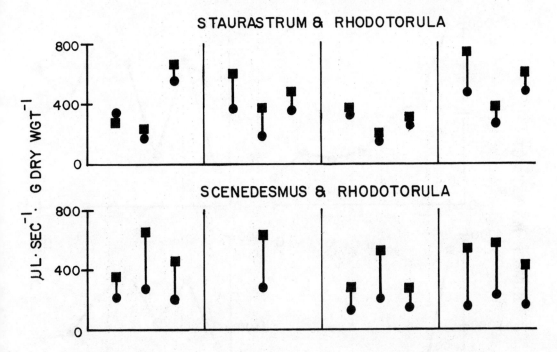

Fig. 2. Clearance rates recorded simultaneously on algae (circles)
and yeast (squares). Each pair of points represents determinations
made on the same portion of sponge. The pairs of points within an
area represent replicate samples from one sponge. Four sponges
were used for each alga.

Scenedesmus – 1.014; t = 25.1; df = 9; p < .001). In addition the slopes of the
two regression lines are significantly different (t = 27.9; df = 20; p < .001)
indicating the differences in the relationship between clearance rates on yeast
and the two algae.

The discrepancies between clearance rates on yeast and Scenedesmus sp occur due
to physiological differences in the sponges' handling of the two particle types.
These differences could occur in the initial uptake or in differential cycling
by the digestive system. The latter caused differential clearance of a bacteria
over yeast by S. lacustris when yeast was present at very high densities (Frost,
in preparation). This explanation seems unlikely in this case. Digestive
handling by the sponge of Scenedesmus sp appears to occur more slowly than for
R. glutinis (see below) under the particle concentrations present in these
experiments. The different size of the two particle types suggested a possible
mechanism for differential uptake. Scenedesmus sp is sufficiently large to be
unable to pass through the ostia of the sponge whereas R. glutinis would traverse
them to be taken up in the incurrent canals and the choanocytes. Uptake from
the surface of the sponge would probably be less than in the interior due to
lower cell densities in this area. I observed, in living sponges, that Scenedesmus sp did occlude the ostia of sponges. However, I also found the same
phenomenon occurring for Staurastrum sp. Usually the ostia of the Valencia
sponge were 35 µm in diameter but I observed some as large as 50 µm and others as
small as 15 µm. Harrison (1972) describes the contractile nature of the porocytes surrounding the ostia. My observations suggest that the diameter of the
ostia may vary with time.

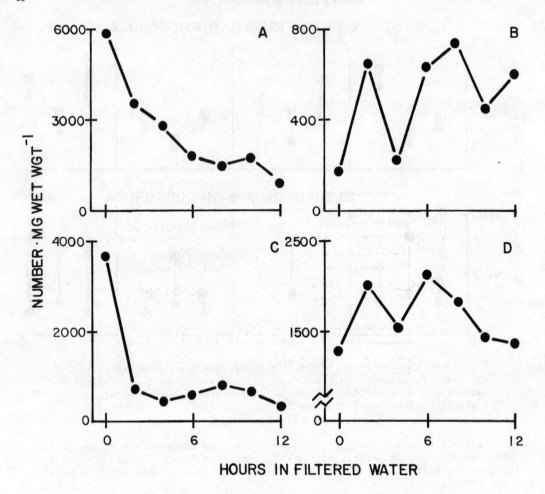

Fig. 3. Concentrations of algal forms within sponges after
varying amounts of time in filtered water; A. Lyngbya limnetica
(number of filaments), B. Nitzschia gracilis, C. Spirulina laxissima,
and D. Nitzschia spp smaller than 40 μm in length.
Note different scales for form number.

Material Handling

Figure 3 shows the effect of sponge digestive processes on 4 particle types. For
two bluegreen algae dramatic reductions occur in particle concentration during
the experiments. In contrast two diatom forms show highly varied concentrations
throughout with no detectable changes. It appears that sponge digestive
processes may have slowed after 4 to 6 hours. However the rates of handling
for these 4 particles during the experiment appear to be quite different.

One potential problem in this analysis occurs in the use of soft algal forms. If
the bluegreens were digested to the point where they could not be identified a
false level of removal would be recorded. To eliminate this possibility I
analyzed sponges for variation in the handling of three diatom forms. In this
experiment I counted only the relative abundance of cells and did not quantify
on a per sponge weight basis. In conjunction with this experiment I tested the

clearance of sponges after 4 hours in filtered water (see Y2 and Y3 above) to ascertain if they were still physiologically active.

The proportion of diatom forms in sponges changed significantly after 4 hours without input (Fig. 4). I tested variation using a G-test and found significant differences in the proportion of cells in 4 control sponges sampled at the start of the experiment (G = 20.78, df = 6; p < .005) and between the groups of 4 control sponges and 4 experimental sponges (G = 7.08; df = 2, p < .005). The proportion of the diatoms in 4 experimental sponges was homogeneous (G = 3.43, df = 6, p < .5). During the experiment the proportion of larger forms,

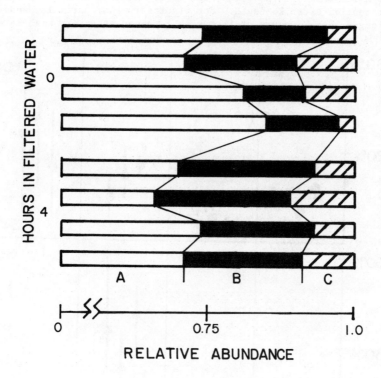

Fig. 4. Relative abundance of three diatom forms in sponges after 0 or 4 hours in filtered water; A. <u>Nitzschia</u> spp smaller than 40 μm in length, B. <u>Nitzschia gracilis</u>, and C. <u>Cyclotella meneghiniana</u>. Note broken scale.

<u>Cyclotella meneghiniana</u> and <u>Nitzschia gracilis</u> increased relative to the smaller <u>Nitzschia</u> spp group. This indicates that the latter were cycled through the digestive system of the sponges at a faster rate than for the larger forms.

Further evidence for differential cycling of cell types is given by the varied amounts of filtered material remaining 12 and 24 hours after feeding on different particle types (see Fig. 5 and discussion below). Considering the proportion of mean values, for three sponges, of material originally taken up, remaining after 12 hours and remaining after 24 hours, the differences are highly significant (G = 31.72 df = 6; p < .001). These differences do not reflect only more

efficient digestion of some forms. This is indicated by the differences in the
proportion retained at the 12 and 24 hour time points.

Retention of Filtered Materials

The amount of material retained by sponges after uptake (Fig. 5) varied significantly with particle type at both 12 (F = 11.3; df = 3,8; p < .001) and 24 hours
(F = 16.76; df = 3,8; p < .001). For R. glutinis and Staurastrum sp no significant change occurred between 12 and 24 hours whereas the proportions of
Ankistrodesmus sp and Scenedesmus sp during the two periods were significantly
different. Aside from Staurastrum sp the particles were reduced to fairly low
levels of retention after 24 hours. For Staurastrum sp, particle handling may
take much longer than 24 hours or it may be retained at a very high level.

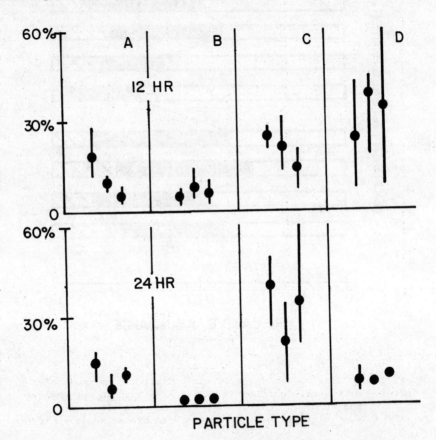

Fig. 5. Proportion of ingested labelled material remaining after
12 and 24 hours; A. Rhodotorula glutinis, B. Ankistrodesmus sp,
C. Staurastrum sp, and D. Scenedesmus sp. Each point represents
the mean of 3 subsamples from one sponge and the bars indicate their range.
The lack of a bar indicates that the range was smaller than the symbol.
These values correspond with sponges shown in Fig. 1 (the yeast values
were derived using the left hand group of sponges).

CONCLUSIONS

My hypothesis that clearance rates would be uniform on a variety of particle types clearly warrants rejection. It appears that variation occurs involving both 1 and 2 in the selection framework.

The difference between yeast clearance rates, in Y1 vs. Y2 and Y2 vs. Y3, suggests that overall water pumping varies. Considering the fact that most of the flow through the sponge must pass through the collars of the choanocytes it is difficult to postulate differences in the actual filtering surfaces. Changes in the amount of suspended materials surrounding the sponge are the most likely cause of the variation in clearance activities as evidenced by the suggestively higher rates for sponges after 4 hours in filtered lake water. The lack of a statistically significant change here may reflect a fairly slow response time to changes in external food concentrations.

The differences in clearance of Staurastrum sp in two experiments may reflect a different mechanism for variation. The size of Staurastrum sp (30 µm in length) is such that it should pass through the largest observed ostia (50 µm) in this sponge. In my observations these large ostia were rare and smaller forms which would exclude Staurastrum were common. If under other conditions the majority of ostia were open to the larger size, the Staurastrum cells would pass through. At the same time, Scenedesmus sp might still be too large to pass the larger ostia. Variation of this form would explain similarities between Staurastrum sp and Scenedesmus sp in one experiment and Staurastrum and R. glutinis in another. It appears that Ankistrodesmus sp, despite its long length, usually passes through the ostia. Harrison (1972) has described the contractile behavior of the cells comprising the ostia. However, I am unfamiliar with any reports describing variation in ostial size with time and this area warrants further investigation. If the ostia do vary they could serve in a sense as a pre-filter screen (2) and at the same time as a filter (3). Although some cells which occlude ostia are ultimately phagocytized (van Weel, 1949) others may be removed by water currents or animal activity on the sponge surface and thus excluded from the filtering system.

The values I have recorded here for clearance rates on R. glutinis are similar to values I have found for Spongilla lacustris (Frost, 1978; in preparation).

My hypothesis that sponges handle particles differentially by varying their turnover time through their digestive system is supported fairly strongly. By tailoring a digestive time and environment to individual particles sponges may exhibit higher efficiencies in utilizing a range of materials when compared to organisms with guts which must operate according to some mean characteristic of the materials they have ingested. Such organisms would either allow the passage of some materials undigested or be forced to process some materials longer than necessary. The sponge's system also permits the rapid turnover of non-desirable particles while maintaining co-occurring materials at normal rates (Frost, in preparation). This system is probably responsible for most short term selection by the sponge, that is, in differential handling of particles at a given time. However, variations in clearance activities could significantly effect the actual ration of materials the sponges utilize particularly when summing effects over extended periods of days or weeks. Overall, selection is therefore a dual level system accomplished by the summation of these two processes. It is not possible to state, given my data, the relative importance of these two systems.

My final hypothesis regarding a uniformity in the efficiency of digestion is difficult to assess. Considerable variation did occur in the amount of ingested material which was retained by the sponges. However, relative similarities did

occur between Scenedesmus sp and yeast which have an outer layer which appears
quite resistant. The ultimate retention of these particles was approached at
different rates. Information on retention of these materials by a number of
organisms in Lago de Valencia is being collected and this will allow at least a
relative consideration of this hypothesis.

Considering the selection framework that I outlined above, variation in sponge
feeding activity occurred chiefly in 2 of the 4 appropriate levels. Water
transport (1) appeared to vary on a relatively long time scale probably in
response to mean characteristics of the suspended materials in its incurrent
stream. Selective digestive activities seemed to operate continuously at an
individual particle level. Also, variation in the size of ostia may be important
as a pre-filtering process (2). Selection at the filtering surfaces (3) which is
common among many suspension feeders (e.g. Lehman, 1976; Donaghay and Small,
1979) does not appear to be important for sponges. The main value of this
selection framework will be in comparisons among organisms. Still, its use in
this case provides valuable insights into sponge feeding processes.

ACKNOWLEDGEMENT

This work was supported by National Science Foundation Grant DEB 7805324 to Wil-
liam M. Lewis, Jr. as part of a joint Venezuelan-American ecosystem study of
Lake Valencia, Venezuela. I thank German Camejo, Susan Knight, Carlos Sevcik,
and Franz Weibezahn for logistic support during this research.

REFERENCES

Bogden, K.G. (1976). The relative abundances and filter-feeding behavior of
    zooplankton: cues to coexistence in the pelagic environment. Ph.D.
    Dissertation, State University of New York.
Donaghay, P.L., and L.F. Small (1979). Mar. Biol., 52, 137-146.
Fjerdingstad, J. (1961). Z. Zellforsch., 53, 645-657.
Frost, T.M. (1976). Sponge feeding: a review with a discussion of some continu-
    ing research p. 283-298 Aspects of Sponge Biology. Academic Press, New York.
Frost, T.M. (1978). Limnol. Oceanogr., 23, 1034-1039.
Harrison, F.W. (1972). Hydrobiologia, 40, 513-517.
Horrocks, D.L. (1974). Applications of liquid scintillation counting. Academic
    Press, New York.
Lewis, W.M. Jr., and F.H. Weibezahn (1976). Arch. Hydrobiol./Suppl., 50, 145-
    207.
Lehman, J.T. (1976). Limnol. Oceanogr. 21, 501-516.
Penney, J.T., and A.A. Racek (1968). Bull. U.S. Nat. Mus., 272, 184 p.
Porter, K.G. (1975). Verh. Int. Verein. Limnol., 19, 2840-2850.
Rasmont, R. and others (1958). Nature, 181, 58-59.
Reiswig, H.M. (1971a). Mar. Biol., 9, 38-50.
Reiswig, H.M. (1971b). Biol. Bull., 141, 568-591.
Ricker, W.E. (1973). J. Fish. Res. Board Can., 30, 409-434.
Sokal, R.R., and F.J. Rohlf (1969). Biometry. W.H. Freeman and Co., San
    Francisco.
Starkweather, P.L., and J.J. Gilbert (1977). Oecologia, 28, 133-139.
van Weel, P.B. (1949). Physiol. Comp., 1, 110-128.

NUTRITION IN SYMBIOTIC TURBELLARIA

J. B. JENNINGS

Department of Pure and Applied Zoology, University of Leeds,
Leeds LS2 9JT, England

ABSTRACT

Patterns of nutritional physiology in symbiotic and free-living Turbellaria are described and compared. Differences in diet, feeding mechanisms, gut structure, digestion and food reserves are shown to be related to the type of host organism favoured by the symbiotes and to the particular site colonized upon or within the host. The various degrees of metabolic dependence on the host, occurring in symbiotic Turbellaria, illustrate possible stages in the past evolution of obligate entoparasitic Platyhelminthes such as the Digenea and Cestoda.

KEYWORDS

Digestion; food reserves; nutrition; parasitism; Platyhelminthes; 'r' and 'K' selection; symbiosis; Turbellaria.

INTRODUCTION

The Turbellaria are predominantly free-living predators but a significant minority (approximately 130 species from at least 27 families) live symbiotically with a wide variety of invertebrate hosts (Hyman, 1951; Jennings, 1971; 1974a). The symbioses range from simple facultative shelter associations, through phoreses and facultative or obligate ecto- and entocommensalism to obligate entoparasitism. All except the first of these types of associations have a strong nutritional basis; the participating Turbellaria, consequently, show various degrees of modification in their nutritional physiology when compared with the basic pattern characteristic of free-living predatory species. These modifications consist of changes in the diets, feeding mechanisms, gut structure, digestive physiology and food reserves of the symbiotes; they reflect increasing degrees of metabolic dependence upon the host and reach a climax in those few turbellarian species that live and feed like cestodes. Thus the different levels of symbiosis apparent in modern living Turbellaria illustrate probable stages in the past evolution of obligate entosymbiotes such as the entoparasitic Trematoda Digenea and Cestoda (Jennings, 1977).

The taxonomy, structure, life histories, geographical distribution
and general physiology of the known symbiotic Turbellaria have been
discussed elsewhere (Jennings, 1971; 1974a; 1977). Consequently the
present account will be restricted to a review of the nutritional
aspects of those few turbellarian symbioses in which these have been
studied in detail. The symbioses selected, which are representative
of the major types of associations found in the Turbellaria, are
summarized in Table I together with some relevant taxonomic information.

NUTRITION IN ECTOSYMBIOTIC TURBELLARIA

Two groups of turbellarian ectosymbioses have been examined from a
nutritional view, namely those between rhabdocoel temnocephalids and
decapod crustaceans, and triclad bdellourids with xiphosuran arthropods (Table I).

Temnocephalid-Decapod Symbioses

Examples studied in detail include <u>Temnocephala novae-zealandiae</u> on
<u>Paranephrops neo-zeylandicus</u> (Fyfe, 1942; Jennings, 1968a) <u>T. bres-
slaui</u> on <u>Aeglea laevis</u> (Gonzales, 1949) and <u>T. brenesi</u> on <u>Macrobra-
chium americanum</u> (Jennings, 1968a).

The temnocephalid rhabdocoels live most commonly on the gills, inner
surfaces of the branchial chambers and general body surfaces of freshwater decapod crustaceans, but a few species occur externally on
freshwater medusae, gastropods and turtles (Jennings, 1971). They
feed principally on diatoms, protozoa, rotifers, nematodes and small
oligochaete annelids and crustaceans and thus have the same diet as
many free-living rhabdocoels (Jennings, 1968b). The prey is captured
by means of tentacles, which is an unusual feature, but its ingestion
by a bulbous doliiform pharynx and subsequent digestion are virtually
identical with these processes as they occur in free-living species.
The digestive physiology, especially, follows the pattern characteristic of the free-living predatory Turbellaria (Jennings, 1974b).
Briefly, endopeptidases from gland cells in the mono-layered gastrodermis initiate extracellular proteolysis; the resulting partially-
disintegrated food is phagocytosed by gastrodermal phagocytes and
digestion is completed intracellularly in a classic food vacuole
(=phagosome)/lysosome/heterolysosome sequence. The enzymes involved
in this have been identified and localised histochemically; they include acid and alkaline phosphatases, endopeptidases, exopeptidases
of the arylamidase "leucine aminopeptidase" type, lipases and amy-
lases.

Food reserves in the Temnocephalida similarly follow the pattern of
free-living species, with lipids stored in the gastrodermis forming
the principal long-term reserve; glycogen occurring at the same sites
but in much smaller amounts forms only a day-to-day reserve and is
rapidly depleted during starvation.

The presence of temnocephalids seems to have little, if any, adverse

TABLE I   Some Typical Turbellarian Symbioses

Turbellaria	Host : Site	Food	Digestion	Food Reserves
Ectosymbioses				
Rhabdocoela : Temnocephalida				
Temnocephalidae e.g. Temnocephala spp.	f.w. Decapoda: gills, branchial chamber	small invertebrates (+ host's food?)	as free-living spp.	as free-living spp., mainly lipids
Tricladida : Maricola				
Bdellouridae e.g. Bdelloura sp.	Xiphosura: gills, ventral surface	other invertebrates host's food	"	" (but less lipids)
Entosymbioses				
Rhabdocoela : Dalyellioida				
Umagillidae e.g. Syndesmis spp.	Echinoidea: gut, coelom	co-symbiotes; host cells (rare)	"	mainly glycogen (18% dry wt.)
Graffillidae e.g. Graffilla sp., Paravortex spp.	Gastropoda; Bivalvia: gut digestive gland	Host's; host cells; "residual bodies"	modified, depends on host's enzymes	mainly glycogen
Fecampiidae e.g. Fecampia spp., Kronborgia spp.	Decapoda, Amphipoda: haemocoel	soluble nutrients in blood	none in classical sense	not known

effect upon the host. Egg-laying causes a build-up of detritus and
encrusting organisms such as protozoa around the egg-cases which are
cemented to the gills, but this is rapidly removed by the newly-
hatched flatworms who take their first meals from this readily available source (Jennings, 1968a). There is a little evidence that stray
fragments of the host's food are utilized, so there may be some opportunistic ectocommensalism. Basically, though, the relationship
seems to be phoretic, with the host merely acting as a feeding plat-
form and providing a current of water which brings the type of food
organisms utilized by free-living rhabdocoels within reach of the
temnocephalids.

Bdellourid-Xiphosuran Symbioses

The only example studied in detail of this type of turbellarian ecto-
symbiosis is that between the triclad bdellourid Bdelloura candida
and the horse-shoe crab Limulus polyphemus (Jennings, 1968a; 1977;
Davis and Fried, 1977; Lauer and Fried, 1977). Here, there is sound
evidence of opportunistic commensalism supplementing the normal triclad diet of small invertebrates such as annelids and crustaceans.
The host shreds its food prior to ingestion and stray particles have
been shown to be swept in the respiratory current across the gills
and ventral body surfaces which the triclad symbiote inhabits. These
particles are captured, and ingested by the tubular plicate pharynx,
in a manner identical with that seen in free-living triclads. Subsequent digestive processes, also, are identical with those of free-
living species (Jennings, 1974b) and those described above for the
ectosymbiotic temnocephalids.

Food reserves, too, broadly resemble those of free-living species and
temnocephalids, but the amount of lipid stored is slightly reduced
(Calow and Jennings, 1974). This is the beginning of a trend which
has become a dominant feature in all entosymbiotic flatworms.

Infestation of Limulus spp. by bdellourids such as B. candida, Syncoelidium pellucidum (Wheeler, 1894) and Ectoplana sp. (Kaburaki,
1922) appears to have no adverse effect, even when as many as 300 are
present on a single host (Jennings, 1974a).

NUTRITION IN ENTOSYMBIOTIC TURBELLARIA

Nutritional aspects of the host/symbiote relationship have been studied in three types of turbellarian entosymbioses. These are, respectively, those between umagillid rhabdocoels and echinoids, graffillid
rhabdocoels and gastropod or bivalve molluscs, and (to a much smaller
extent) fecampiid rhabdocoels and amphipod or decapod crustaceans
(Table I).

Umagillid-Echinoid Symbioses

Examples of entosymbiotic umagillids studied include Syndesmis antill-
larum, living in the gut and perivisceral space of Lytechinus varie-
gatus, and S. franciscana from the same sites in Strongylocentrotus

franciscanus and S. purpuratus (Jennings and Mettrick, 1968; Mettrick
and Jennings, 1969). Both species feed on the bacteria and ciliate
protozoa which occur in very large numbers at these sites within the
hosts; the nutritional physiology of the co-symbiotes is unknown but
they do not appear to be competing for the host's nutrients since infected echinoids show no ill-effects attributable to their presence.
It may be that the three groups of symbiotic organisms form a simple
food-chain, with the autotrophic bacteria as "primary producers"
which are then utilized either directly by the umagillids or indirectly via the protozoa.

There is some evidence that S. antillarum occasionally ingests host
coelomocytes along with the ciliates but the small numbers found in
the gut contents suggest that this is probably by chance and that
host cells and tissues, therefore, do not form a significant part of
the diet.

The digestive physiology in both species of Syndesmis is virtually
identical with that found in free-living rhabdocoels, as would be
expected from the diet and gut structure which are very similar to
those of free-living species such as Stenostomum and Macrostomum.

There is, however, one outstanding difference between these umagillids and the free-living flatworms. The food reserves show a marked
shift in emphasis from lipid storage, which is a characteristic
feature of most free-living predators, towards deposition of glycogen.
In S. franciscana, for example, glycogen constitutes approximately
18% of the dry weight, lipid reserves are negligible and there is
consequently a major shift in calorific value when compared with
free-living rhabdocoels (Calow and Jennings, 1974).

Graffilid-Mollusc Symbioses

Three entosymbiotic species from the rhabdocoel family Graffillidae
have been investigated. These are Paravortex scrobiculariae, living
in the intestine and digestive gland of the bivalve Scrobicularia
plana; Paravortex cardii from the main ducts of the digestive gland
in the bivalve Cerastoderma edule, and Graffilla buccinicola from the
digestive gland of the gastropods Buccinum undatum and Neptunea
decemcostata (Jennings and Phillips, 1978).

In all three species the diet, gut structure, digestive physiology
and food reserves are much modified relative to those characteristic
of the free-living and ectosymbiotic Turbellaria; they show adaptive
features related to the entosymbiotic habit and, in particular, to
the feeding and digestive processes of their respective hosts. All
feed on semi-digested components of their hosts' diets, supplemented
by a significant amount of cellular debris from the hosts' digestive
glands. This debris contains host digestive enzymes which play a
major role in the continuation, extension and completion of digestion within the rhabdocoels' own alimentary systems. The utilization of host enzymes by the symbiotes has apparently resulted in the
loss of specific enzyme-producing gland cells from their own intes-

tinal gastrodermis; this interpretation of the absence of morphologically distinct gland cells is supported by the fact that the closely
related, but free-living, Pseudograffilla arenicola has a normal gastrodermis fully differentiated into glandular and phagocytic components (Jennings, unpublished data).

Utilization of the host's cells and enzymes, in addition to its
partly digested food, is facilitated by certain aspects of the hosts'
digestive physiology which are peculiar to the bivalve and gastropod
molluscs. These have been reviewed by Owen (1966; 1974), Purchon
(1977) and Jennings and Phillips (1970). In brief, preliminary extracellular digestion in the molluscan stomach is followed by movement
of food into the digestive gland where it is phagocytosed and digestion completed intracellularly. A proportion of the digestive cells
then disintegrates and cellular debris, "residual bodies" (spent
heterolysosomes) and some active enzymes pass back down the ducts of
the digestive gland to the stomach. Here the enzymes help initiate
digestion of the next meal (supplementing those of the crystalline
style in bivalves) and the remainder of the discarded material is
passed into the intestine for elimination in the faeces. In the
inter-tidal bivalves feeding, gastric digestion and the events within
the digestive gland are intimately related to, and cued by, tidal
factors. The length of time between successive submergences, particularly, has a strong influence and in species living at or near
high-water mark, such as Scrobicularia plana (the host of Paravortex
scrobiculariae) feeding and digestion occur in synchronised clear
cut diphasic cycles.

The entosymbiotic graffillids, lying within the ducts of the digestive gland, abstract their hosts' partly digested food as this passes
up the ducts into the gland, and cellular debris and enzymes as these
pass in the reverse direction. The three species show some variation
in their locations within the host's gut and this is related to the
host's cycles of feeding and digestion and, therefore, to the type
of habitat favoured by the host and the length of time between con-
secutive tidal submergences. Graffilla buccinicola, in the sub-
littoral Buccinum undatum and Neptunea decemcostata, lives permanently in the upper parts of the digestive gland and sometimes actually
within the distal digestive tubule proper, and on occasion supple-
ments its diet by actively removing intact digestive cells from the
epithelium. Paravortex cardii, in Cerastoderma edule inhabiting sand
around low-water mark, lives permanently in the lower digestive gland
in the main (primary) ducts. In contrast, P. scrobiculariae spends
part of its time in the host's anterior intestine and migrates into
the lower digestive gland only to feed. Its host, living in mud near
high-water mark, has to endure long periods (8 - 10 hours) between
tidal submergences and during that time the oxygen content of the
mud declines rapidly. The feeding migrations of P. scrobiculariae
are synchronized with its host's cycles of feeding and digestion,
the symbiote remaining in the intestine at high-water when the host
is feeding but migrating into the digestive gland on the ebb-tide
when partly digested food is passing up into the gland from the
stomach. It remains in the gland well into the period of cell disintegration, ingests the resultant cellular debris passing back to
the stomach and then returns to the intestine on the flood-tide
before the host re-commences feeding.

P. scrobiculariae is thus most active over the period when oxygen
supplies in the host's habitat of mud are likely to be minimal. It
shows two features related to this situation, which make it the most
highly adapted of the entosymbiotic graffillids. A physiologically
active haemoglobin occurs in and around the brain and pharynx and
this is believed to make such oxygen as is present, in the habitat
and host tissues, more easily available to the symbiote by a process
of facilitated diffusion (Phillips, 1978). Secondly, the symbiote
is apparently capable of releasing energy by anaerobic glycolysis, as
well as by the normal aerobic pathways; dehydrogenases concerned with
glycolysis occurring around the pharynx, brain and vitellaria in much
larger amounts in this species than in the other symbiotic graffillids (Jennings and LeFlore, 1979).

Fecampiid-Crustacean Symbioses

The family Fecampiidae contains the three genera Fecampia, Glandulo-
derma and Kronborgia; virtually all the known species live in the
body cavity of various isopod and decapod crustaceans (Jennings, 1971).
The exceptional species, Fecampia balanicola, inhabits the mantle
cavity, ovaries and opercular muscles of cirripedes (Christensen and
Hurley, 1977). The best known examples, in terms of structure and
life history are F. erythrocephala from the haemocoel of decapods
such as Cancer pagurus, Pagurus bernhardus and Carcinus maenas (Giard,
1886; Brun 1967), F. xanthocephala (regarded by Christensen and
Hurley, 1977, as the same species as F. erythrocephala) from the isopod Idotea neglecta (Caullery and Mesnil, 1903), and the amphipods
Gammarus locusta and G. duebeni (Toulmond and Truchot, 1964), Kronborgia amphipodicola from the haemocoel of the amphipods Ampelisca
macrocephala and A. tenuicornis (Christensen and Kanneworff, 1964;
1965), and K. caridicola from the shrimps Eualus machilenta, Lebbeus
polaris and Paciphaea tarda (Kanneworff and Christensen, 1966).

The range of hosts recorded for most of these species is unusually
broad and this lack of host specificity, whilst particularly evident
in the fecampiids, occurs throughout the symbiotic Turbellaria, where
certain taxa are always associated with similar groupings of host
species (Jennings, 1974a). It is seen also, of course, among the
Trematoda Monogenea, the Digenea (to a lesser extent) and the Cestoda.

None of the fecampiid symbioses have been investigated specifically
from nutritional aspects but enough information can be gleaned from
these other accounts to allow deductions regarding the mode of nutrition and the degree of dependence on the host. A feature common
to all described species is the absence of a functional alimentary
system from the adult flatworms, although various parts may be present in earlier stages. In Fecampia erythrocephala/xanthocephala,
for example, the young superficially resemble free-living rhabdocoels and possess eyes, mouth, "buccal tube", pharynx and intestine.
After entering the host's haemocoel, however, they quickly lose the
eyes, mouth, "buccal tube" and pharynx so that, while the intestine
persists in most cases as a closed sac, normal ingestion and digestion of particulate food obviously cannot occur. The flatworm grows,
becoming sexually mature, and must, therefore, be obtaining nutrients
in soluble form from the host's haemocoelic fluid. The hosts show

no ill-effects, indicating that the relationship is an ancient one to
which both partners are fully adapted.

A slightly different situation is found in the <u>Kronborgia amphipodi-
cola</u>/amphipod symbiosis. Here eyes, mouth, pharynx and intestine are
absent from all stages in the life cycle so that epidermal uptake of
nutrients must occur throughout life, once the reserves from the egg-
yolk have been used. The epidermis shows some structural modifications for this process, in that the microvilli which occur between
the cilia are longer and more numerous than those in free-living
flatworms (Bresciani and Køie, 1970). This presumably increases the
surface area available for absorption; a further possibility is that
the microvilli are also concerned with membrane or "contact" digestion of components of the haemocoelic fluid, in the same way as those
of the vertebrate intestine are concerned with "contact" digestion of
carbohydrates (Ugolev, 1965).

Abstraction of nutrients from the haemocoelic fluid by <u>Kronborgia</u> has
a definite and deleterious effect upon the host; both male and female
hosts show progressive atrophy of the gonads as the flatworms grow
and become sexually mature, presumably due to the symbiotes' successful competition for available nutrients, and finally the hosts become
sterile. <u>Kronborgia</u> eventually leaves the host, reproduces and dies;
its migration through the body wall and exoskeleton kills the host
but the direct cause of its death remains unknown.

<u>Kronborgia</u> would thus appear to be less well adapted to its mode of
life than are the other entosymbiotic rhabdocoels, in the sense that
it prohibits its hosts from breeding and eventually causes their
death. From purely nutritional aspects, though, it is one of the
most modified turbellarian symbiotes; the loss of the gut from all
stages in the life history is the logical termination of the process
which begins in the graffillids with the utilization of, and dependence on, host digestive enzymes.

Nothing is known of food reserves in the Fecampiidae.

DISCUSSION

The five types of turbellarian symbioses described here show increasing degrees of metabolic dependence of the symbiotes upon their
hosts. They illustrate, therefore, possible stages in the past evolution of entosymbiotes such as modern cestodes which are now wholly
dependent, nutritionally, on host organisms.

The degree of dependence, and thus the amount of modification away
from the nutritional pattern characteristic of free-living predatory
Turbellaria, is directly related to the type of host animal and the
site upon or within it colonised by the symbiote. These factors
determine the type of food available and from this spring all subsequent nutritional changes. Thus, the ectosymbiotic temnocephalids
feed and digest in a manner identical with that of free-living

rhabdocoels; the relationship with their decapod hosts is phoretic
with perhaps some opportunistic commensalism and the flatworms can
survive and even breed <u>in vitro</u> away from their hosts (Gonzales, 1949;
Hickman, 1967; Jennings, 1968c). The ectosymbiotic bdellourids similarly retain a nutritional pattern akin to that of their free-living
predatory relatives but the coarse feeding habits of the host <u>Limulus</u>
permits a larger and permanent opportunistic commensal component.

Of the entosymbiotic species the umagillids are the equivalent,
nutritionally, of the temnocephalids in that their diet and digestive
physiology differs little from free-living forms in spite of the very
different habitat. This is due, of course, to the extreme abundance
of co-symbiotic bacteria and protozoa; these are a characteristic
feature of echinoids (Hyman, 1955) and so the <u>type</u> of host, in this
case, has a dominating influence.

An outstanding and highly significant feature of the umagillid/echin-
oid symbiosis, though, is the increased emphasis on glycogen storage
by the turbellarians. This is a feature of all entosymbiotic flatworms, whatever their metabolic relationship with their hosts, and
has been linked, in terms of 'r' and 'K' selection, with the increased
fecundity which is similarly a feature of all entosymbiotes. Such
increased fecundity can therefore be regarded as a direct <u>consequence</u>
of the readily available and constant food supply provided by the
entosymbiotic habit, rather than as a basic prerequisite for the latter life style (Calow and Jennings, 1974; Jennings and Calow, 1975).

The graffillid entosymbioses, like those involving the umagillids,
are influenced by the type of host selected, but here it is the particular nature of the hosts' own digestive physiology (linked with
the persistence in molluscs of extensive intracellular digestion)
which has had a profound effect upon the symbiotes. Utilization of
host digestive enzymes, ingested along with the food, has apparently
allowed loss of many endogenous enzymes. The molluscan digestive
gland would seem to be an excellent entosymbiotic habitat and it may
well have been the primitive habitat of those earlier symbiotic rhabdocoels which eventually gave rise to digenetic trematodes; it certainly remains the favoured habitat of very many modern larval Digenea. Where the mollusc host itself lives under fairly extreme conditions, which in turn affect the symbiotes, the latter have developed
further adaptive features. <u>Paravortex scrobiculariae</u>, for example,
has apparently evolved a haemoglobin and an increased emphasis on
anaerobic glycolytic respiration as adaptations permitting it to syn-
chronize its own feeding activities with the appropriate parts of its
host's tidally cued feeding and digestive cycles (Phillips, 1978;
Jennings and LeFlore, 1979). The capability for glycolysis, apart
from its intrinsic interest in this single species where it is of
only intermittent, albeit regular, use, is also extremely relevant in
connection with the evolution of entosymbiotes such as the Digenea
and Cestoda. In these, glycolysis and related types of intermediary
metabolism often play a dominant part in the life processes and permit survival in permanently unfavourable circumstances.

The fecampiid/crustacean symbioses would obviously repay further investigation. It is apparent from such studies as are available that

Fecampia and Kronborgia represent the climax to the evolution of turbellarian entosymbioses and apparently parallel the cestodes in many
features such as the absence of a functional gut and the development
of tegumentary absorption of nutrients. Biochemical aspects of their
nutrition, however, remain unknown.

SUMMARY

1. Symbiotic Turbellaria show a range of adaptive modifications in
their nutritional physiology related to the type of host organism,
their particular location upon or within the host, and the type of
food available to them.

2. Temnocephalid rhabdocoels and bdellourid triclads (ectosymbiotic
on arthropods) and umagillid rhabdocoels (entosymbiotic in echino-
derms) have available to them food organisms of the same type as are
utilized by free-living flatworms and show no significant differences
from these in feeding mechanisms or digestion.

3. Graffillid rhabdocoels (entosymbiotic in molluscs) feed on their
host's partly digested food, digestive gland debris and enzymes. The
digestive physiology is much modified and the host's digestive enzymes
play a dominant role in the symbiotes' alimentary systems.

4. Fecampiid rhabdocoels (entosymbiotic in crustaceans) absorb soluble nutrients from the host's haemocoelic fluid across their body
wall and do not have a functional alimentary system in the adult.

5. Entosymbiotic species, irrespective of dietary or digestive modifications, store glycogen rather than lipid. This is the reverse of
the situation in free-living species and is linked with the entosymbiotic habit, the assured food supply and high fecundity.

6. Modifications in nutritional physiology related to life style in
the symbiotic Turbellaria illustrate possible stages in the evolution
of other entosymbiotic Platyhelminthes such as the Trematoda Digenea
and the Cestoda.

REFERENCES

Bresciani, J., and M. Køie (1970). On the ultrastructure of the epi-
    dermis of the adult female of Kronborgia amphipodicola Christensen
    and Kanneworff, 1964 (Turbellaria, Neorhabdocoela) Ophelia, 8, 209-
    230.
Brun, B. (1967). Sur la presénce de turbellariés parasites du genre
    Fecampia en Mediterranée. Bull. Mus. Hist. Nat. Marseille, 27,
    141-145.
Calow, P., and J. B. Jennings (1974). Calorific values in the phylum
    Platyhelminthes: the relationship between potential energy, mode
    of life and the evolution of entoparasitism. Biol. Bull., 147,
    81-94.

Caullery, M., and F. Mesnil (1903). Recherches sur les "Fecampia" Giard, Turbellariés Rhabdocéles, parasites internes des crustacés. Ann. Fac. Sci. Marseille, 13, 131-168.

Christensen, A. M., and A. C. Hurley (1977). Fecampia balanicola sp. nov. (Turbellaria Rhabdocoela), a parasite of Californian barnacles. Pages 119-128 in T. G. Karling and M. Meinander (Eds.), The Alex. Luther Centennial Symposium on Turbellaria. Acta Zoologica Fennica, 154, Societas pro Fauna et Flora Fennica, Helsinki.

Christensen, A. M., and B. Kanneworff (1964). Kronborgia amphipodicola gen. et sp. nov., a dioecious turbellarian parasitizing ampeliscid amphipods. Ophelia, 1, 147-166.

Christensen, A. M., and B. Kanneworff (1965). Life history and biology of Kronborgia amphipodicola Christensen and Kanneworff (Turbellaria, Neorhabdocoela). Ophelia, 2, 237-251.

Davis, R. E., and B. Fried (1977). Histological and histochemical observations on Bdelloura candida (Turbellaria) maintained in vitro. Trans. Amer. Microsc. Soc., 96, 258-263.

Fyfe, M. L. (1942). The anatomy and systematic position of Temnocephala novae-zealandiae Haswell. Trans. R. Soc. N. Z., 72, 253-267.

Giard, M. A. (1886). Sur une rhabdocoele nouveau, parasite et nidulant (Fecampia erythrocephala). C. R. Acad. Sci. Paris, 103, 499-501.

Gonzales, M. D. P. (1949). Sôbre a digestao e a respiraçâo des Temnocephalas (Temnocephalus bresslaui spec. nov.). Bolm. Fac. Filos. Ciênc. Univ. S. Paulo (Zool.), 14, 277-323.

Hickman, V. V. (1967). Tasmanian Temnocephalidae. Pap. Proc. R. Soc. Tasm., 101, 277-250.

Hyman, L. H. (1951). The Invertebrates: Platyhelminthes and Rhynchocoela. The acoelomate Bilateria. Volume II. McGraw-Hill Book Co., New York.

Hyman, L. H. (1955). The Invertebrates: Echinodermata. The coelomate Bilateria. Volume IV. McGraw-Hill Book Co., New York.

Jennings, J. B. (1968a). Feeding, digestion and food storage in two species of temnocephalid flatworms (Turbellaria: Rhabdocoela). J. Zool. London, 156, 1-8.

Jennings, J. B. (1968b). Nutrition and digestion. Pages 303-326 in M. Florkin and B. T. Scheer, Eds., Chemical Zoology, Volume II, Porifera, Coelenterata and Platyhelminthes. Academic Press, New York.

Jennings, J. B. (1968c). A new temnocephalid flatworm from Costa Rica. J. Nat. Hist., 2, 117-120.

Jennings, J. B. (1971). Parasitism and commensalism in the Turbellaria. Pages 1-32 in B. Dawes, Ed., Advances in Parasitology. Volume IX. Academic Press, New York.

Jennings, J. B. (1974a). Symbioses in the Turbellaria and their implications in studies on the evolution of parasitism. Pages 127-160 in Winona B. Vernberg, Ed., Symbiosis in the Sea. University of South Carolina Press, Columbia, South Carolina.

Jennings, J. B. (1974b). Digestive physiology of the Turbellaria. Pages 173-197 in N. W. Riser and M. P. Morse, Eds., Biology of the Turbellaria, Libbie H. Hyman Memorial Volume. McGraw-Hill Book Co., New York.

Jennings, J. B. (1977). Patterns of nutritional physiology in freeliving and symbiotic Turbellaria and their implications for the evolution of entoparasitism in the phylum Platyhelminthes. Pages 63-79 in T. G. Karling and M. Meinander (Eds.), The Alex. Luther Centennial Symposium on Turbellaria. Acta Zoologica Fennica, 154,

Societas pro Fauna et Flora Fennica, Helsinki.

Jennings, J. B., and P. Calow (1975). The relationship between high fecundity and the evolution of entoparasitism. Oecologia (Berlin), 21, 109-115.

Jennings, J. B., and D. F. Mettrick (1968). Observations on the ecology, morphology and nutrition of the rhabdocoel turbellarian Syndesmis franciscana (Lehman, 1946) in Jamaica. Caribbean J. Sci., 8, 57-69.

Jennings, J. B., and W. B. LeFlore (1979). Occurrence and possible adaptive significance of some histochemically demonstrable dehydrogenases in two entosymbiotic rhabdocoels (Platyhelminthes : Turbellaria). Comp. Biochem. Physiol., 62B, 301-304.

Jennings, J. B., and J. I. Phillips (1978). Feeding and digestion in three entosymbiotic graffillid rhabdocoels from bivalve and gastropod molluscs. Biol. Bull., 155, 542-562.

Kaburaki, T. (1922). On some Japanese Tricladida Maricola, with a note on the classification of the group. J. Coll. Sci. Imp. Univ. Tokyo, 44, 1-54.

Kanneworff, B., and A. M. Christensen (1966). Kronborgia caridicola sp. nov., an endoparasitic turbellarian from North Atlantic shrimps. Ophelia, 3, 65-80.

Lauer, D. M., and B. Fried (1977). Observations on nutrition of Bdelloura candida (Turbellaria), an ectocommensal of Limulus polyphemus (Xiphosura). Amer. Mid. Nat., 97, 240-247.

Mettrick, D. F., and J. B. Jennings (1969). Nutrition and chemical composition of the rhabdocoel turbellarian Syndesmis franciscana, with notes on the taxonomy of S. antillarum. Can. J. Zool., 26, 2669-2679.

Owen, G. (1966). Digestion. Pages 53-96 in K. M. Wilbur and C. M. Yonge, Eds., Physiology of Mollusca, Volume II. Academic Press, New York.

Owen, G. (1974). Feeding and digestion in the Bivalvia. Pages 1-35 in O. Lowenstein, Ed., Advances in Comparative Physiology and Biochemistry, Volume V. Academic Press, New York.

Phillips, J. I. (1978). The occurrence and distribution of haemoglobin in the entosymbiotic rhabdocoel Paravortex scrobiculariae (Graff) (Platyhelminthes: Turbellaria) Comp. Biochem. Physiol., 61C, 679-683.

Purchon, R. D. (1977). The Biology of the Mollusca, 2nd Ed. Pergamon Press, Oxford.

Toulmond, A., and J-P. Truchot (1964). Inventaire de la Faune Marine de Roscoff. Amphipodes, Cumacés. Suppl. Trav. Stat. Biol. Roscoff, 1964.

Ugolev, A. M. (1965). Membrane (contact) digestion. Physiol. Rev., 45, 555-595.

Wheeler, W. M. (1894). Syncoelidium pellucidum, a new marine triclad. J. Morph., 9, 167-194.

SOME EFFECTS OF DIET ON THE BIOLOGY OF THE ROTIFERS *ASPLANCHNA* AND *BRACHIONUS*

J. J. GILBERT

Department of Biological Sciences, Dartmouth College, Hanover,
New Hampshire 03755, U.S.A.

ABSTRACT

In certain rotifers the size, biomass, shape, population dynamics, and certain
life cycle events, such as the occurrence of sexual or parthenogenetic reproduction and the breaking of dormancy, may be controlled by the type of food eaten.

In the A. brightwelli-A. intermedia-A. sieboldi series, there is an increasingly
pronounced developmental polymorphism in female size and shape controlled by two
dietary factors -- tocopherol (vitamin E) and food type. In the absence of tocopherol, females are relatively small, saccate, and amictic (female-producing).
When tocopherol is present, they are 50-200% larger in length, may possess characteristic body-wall outgrowths, and are often mictic (male-producing). In A.
intermedia and A. sieboldi, there are two tocopherol-dependent morphotypes --
cruciform and campanulate. The latter is induced when the diet consists of congeneric or crustacean prey.

Within each morphotype of these species, food type may exert considerable control
over body size and shape, biomass, and life-table parameters. In A. sieboldi,
saccate females fed on Brachionus are larger, have a greater dry weight biomass,
produce more than twice as many offspring, live longer, and can reproduce more
rapidly than those fed on Paramecium. In all morphotypes, but especially the tocopherol-dependent ones, large prey induce larger body sizes and disproportionately
larger coronae than small prey.

Tocopherol-induced body enlargement involves an increase in cell number in certain
cellular structures, an increase in nuclear number, size, and DNA endoreduplication
in some syncytial structures, and additional cytoplasmic growth. The mechanism by
which food type can modify or exaggerate these growth responses is not understood.

In B. calyciflorus, food type can influence body size, relative length of posterolateral spines, and the hatching pattern, as well as the hatchability, of resting
eggs.

KEYWORDS

Rotifer; vitamin E; growth; developmental polymorphism; resting-egg hatching;
dietary effects.

57

INTRODUCTION

The purpose of this paper is to show that the qualitative nature of the diet may influence the morphology, development, physiology, and ecology of some rotifers in many diverse and often dramatic ways. For example, size, shape, biomass, population dynamics, and even life cycle events may be controlled by the type of food eaten. Some of these nutritional effects pose extremely interesting questions about mechanisms of action, and many of them should be considered in analyses of natural populations and before results from laboratory populations cultured on one type of diet are generalized.

The Rotifera and their close relatives, the Gastrotricha, share many features with the Turbellaria and may have evolved from an ancestral, benthic acoel (Beauchamp, 1909; Steinböck, 1958). Some of the dietary effects described in this paper may be restricted to certain rotifers, but others certainly have counterparts in other lower metazoa and in more advanced invertebrate groups as well.

This paper deals with observations in our laboratory on several species of the ovoviviparous predator, Asplanchna, and on the oviparous suspension feeder, Brachionus calyciflorus. The material on Asplanchna is extracted from previously published papers, while that on Brachionus is original. The paper is subdivided into six sections, each with its own experimental methods, results, and discussion.

EXPERIMENTS: METHODS, RESULTS, DISCUSSION

Effect of Food Type on the Size, Biomass, Longevity, Fecundity, and Reproductive Rate of Saccate Asplanchna sieboldi

In the absence of dietary tocopherol (vitamin E), the potentially polymorphic females of this species always belong to the relatively small, saccate morphotype and reproduce exclusively by female parthenogenesis. In the study described here, summarized from Gilbert (1976, 1977a), the two food organisms tested were both tocopherol-free so that transformations to the larger, tocopherol-dependent morphotypes (see below) could be avoided.

Populations were maintained on the ciliate Paramecium aurelia, which had been cultured on the bacterium Aerobacter aerogenes, and on the rotifer Brachionus calyciflorus, which had been cultured on the yeast Rhodotorula glutinis. Asplanchna from populations cultured on these foods for at least several generations were then compared.

Adult A. sieboldi on the two diets were very similar in size, although those fed on Brachionus were slightly but significantly larger in either body length, body width, or head width.

The dry weight biomass of the Brachionus-fed Asplanchna was considerably greater than that of the Paramecium-fed ones, newborn and gravid, adult females of the former being 1.26 and 3.33 μg per female and those of the latter being 0.56 and 2.02 μg per female. Thus, Brachionus-fed animals were 125% and 65% heavier than Paramecium-fed ones as neonates and adults, respectively. These results also show that the postnatal growth of the Brachionus-fed Asplanchna was proportionately considerably less than that of the Paramecium-fed animals, being 2.6 and 1.6 times the birth weight, respectively.

This analysis demonstrates that the biomass of an organism can be greatly influenced by its diet. Even though Paramecium and Brachionus prey can both sustain indefinite reproduction (Gilbert, unpublished), the latter is a markedly superior food. This fact is even more clearly shown below in a comparison of the dynamics of populations cultured on these two food types.

The fecundity, survivorship, and reproductive potential of Brachionus-fed Asplanchna were all dramatically greater than those of Paramecium-fed Asplanchna (Table 1). The former produced 2.1 to 2.3 times as many offspring and lived about twice as long as the latter. The greater survivorship was primarily due to a prolongation of the post-reproductive period. On account of their much greater rate of offspring production, the Brachionus-fed animals had a calculated finite rate of population growth 43-63% higher than that of the Paramecium-fed animals.

TABLE 1   Age (x)-specific Survivorship (1_x) and Fecundity (m_x), Number of Offspring per Female per Generation (R_0) ($R_0 = \Sigma 1_x m_x$), Instantaneous Rate of Population Growth (r)($1 = \Sigma e^{-rx} 1_x m_x$), and Finite Rate of Population Growth (λ) ($\lambda = e^r$) of Saccate Asplanchna sieboldi fed on Paramecium aurelia and on Brachionus calyciflorus at 26°C. 20-21 Asplanchna per Cohort.

Age interval (x) (days) and rate of reproduction	Food type and experiment							
	Paramecium				Brachionus			
	1		2		1		2	
	1_x	m_x	1_x	m_x	1_x	m_x	1_x	m_x
0 - .5	1.00	0	1.00	0	1.00	0	1.00	0
.5 - 1	1.00	1.25	1.00	1.80	1.00	2.85	1.00	2.52
1 - 1.5	1.00	2.05	1.00	2.05	1.00	3.90	1.00	3.91
1.5 - 2	1.00	2.10	1.00	1.70	1.00	3.60	1.00	3.71
2 - 2.5	1.00	1.80	1.00	1.70	1.00	3.50	1.00	2.86
2.5 - 3	.65	1.46	.65	1.54	1.00	2.50	.95	2.60
3 - 3.5	.10	1.50	.20	1.00	1.00	2.00	.95	1.65
3.5 - 4	0	0	0	0	.80	.50	.95	.50
4 - 4.5					.75	0	.95	.15
4.5 - 5					.55	0	.91	0
5 - 5.5					.30	0	.91	0
5.5 - 6					.15	0	.81	0
6 - 6.5					0	0	.62	0
6.5 - 7							.38	0
7 - 7.5							.14	0
7.5 - 8							0	0
R_0	8.30		8.45		18.75		17.67	
r	1.22		1.30		1.71		1.66	
λ	3.39		3.67		5.53		5.26	

These data show that food type can have a striking effect on the population dynamics of a species. There are several obvious ecological implications of such effects. Food type certainly must be considered when attempts are made to relate the population growth of a species in nature to available food material. For

example, a population could have a slow birth rate even when food was abundant if
that food could not be readily assimilated or was low in nutrient quality. This
point was previously made by King (1967), who found that the reproductive potential
of the rotifer Euchlanis dilatata in laboratory populations varied greatly with
the type of alga used as food. Food type also should be considered when competitive interactions between two or more species are being analyzed. Relative competitive abilities are often inferred from laboratory studies in which the reproductive rates of the species in question are determined from cultures on a single
food source. The potential population growth rates of the species, however, may
vary considerably and dissimilarly with food type. Thus, the relative competitive
abilities of the species may shift as a function of food type.

Some work, however, has shown that a variety of food types may be very similar in
their abilities to sustain reproduction. Theilacker and McMaster (1971), for
example, found that the growth rates of laboratory populations of Brachionus
plicatilis were much the same when cultured on four different algae.

Effect of Dietary Tocopherol (Vitamin E) on Body Size and Shape in Asplanchna

Tocopherol is synthesized almost exclusively by photosynthetic organisms, with
relatively very small amounts also being produced by some fungi (Draper, 1970;
Litton and Gilbert, 1975). Asplanchna, therefore, ingests tocopherol whenever it
eats algal material directly or, more likely,preys on animals which have eaten
algae.

In females of three species of Asplanchna, dietary tocopherol induces and permits
growth responses which result in substantial increases (50-200%) in body size and
alterations of body shape. The many studies conducted on these developmental
polymorphisms have been reviewed (Gilbert, 1974a,b, 1977b, in press), and only
certain aspects will be considered below.

By itself, tocopherol may induce a considerable increase in body size and change
in body shape. In all of the studies described in this section, the food organism
used is Paramecium aurelia. Other prey, in combination with tocopherol, may
promote further or somewhat different growth responses, which will be discussed in
the next two sections.

When grown on Paramecium in the absence of tocopherol, the usual adult female body
size of A. brightwelli is 380-440 μm and that of both A. intermedia and A. sieboldi
is 500-670 μm. However, when cultured on Paramecium with 10^{-8} to 10^{-7} M d-α-tocopherol for several generations, the adult female body sizes of all three species
increase by 50-65% (Gilbert, 1975). This size increase is effected during both
embryogenesis and postnatal development (Gilbert, 1974a).

During prenatal growth, tocopherol stimulates certain regions of the hypodermis to
grow more than others, and so the shapes of newborn individuals which develop with
and without tocopherol are very different (Gilbert and Thompson, 1968, Gilbert,
1975). Embryos containing no tocopherol develop into saccate females, while those
containing more than 5-10 X 10^{-16} moles of this compound (Birky and Gilbert, 1972)
develop into females which have characteristic body-wall outgrowths. These outgrowths are slight in A. brightwelli but may be pronounced in A. intermedia and,
especially, A. sieboldi (Fig. 1). In the latter two species, females have four
outgrowths (two lateral, one posterior, and one posterodorsal), giving them a
cross-shaped appearance from the dorsoventral aspect, and are thus called
cruciform. Tocopherol cannot induce such differential hypodermal growth during

postnatal development, and so the shape of an individual is fixed at birth.

FEMALE MORPHOTYPES IN ASPLANCHNA

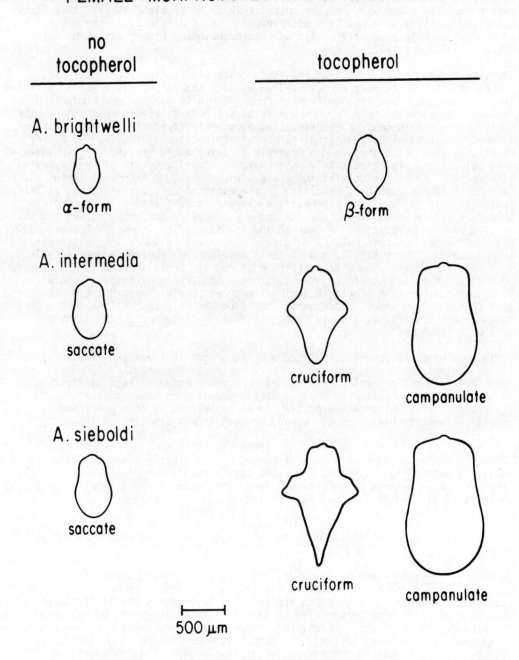

Fig. 1.   Tocopherol-controlled female polymorphism in
three species of Asplanchna

Associated with tocopherol-induced outgrowths is a marked change in reproductive mode (Gilbert and Thompson, 1968). In all three species, saccate females are always amictic, producing diploid eggs which develop parthenogenetically into females, while cruciform females are often mictic, producing haploid eggs which develop parthenogenetically into males. In each species, those females with the most pronounced outgrowths are the most likely to be mictic (Gilbert, 1975; Kabay and Gilbert, 1977). In one laboratory population of A. sieboldi, the probability of cruciforms with maximally developed outgrowths being mictic was about 63% (Kabay and Gilbert, 1977).

Thus, in three species of Asplanchna, the presence of tocopherol in the diet causes an increase in body size, a change in body shape, and, as a result of these two responses, a shift in some females from diploid to haploid egg production. The mechanisms responsible for these effects are only partially understood. The considerable enlargement of the syncytial hypodermis in all tocopherol-dependent morphotypes involves no additional mitoses in this tissue (Birky, 1968; Gilbert et al., in press) and thus is due entirely to cytoplasmic growth. In campanulates (see below), and almost certainly in cruciforms as well, growth of some syncytial structures, such as the vitellarium and gastric glands, is associated with: a) an increase in nuclear number, resulting from additional cell divisions during the early, mitotic phase of embryogenesis; b) larger nuclei; and c) in the latter organs, a higher level of nuclear DNA endoreduplication occurring during the post-mitotic phase of embryogenesis (Wurdak and Gilbert, 1976; Jones and Gilbert, 1977). Both cruciforms and campanulates have many more flame cells on the protonephridia than saccates (Powers, 1912), and so their development involves an increase in cell divisions during the differentiation of certain cellular structures. The production of haploid, male eggs is due to the occurrence of meiotic maturation divisions, rather than single, equational maturation divisions, but the cause of this change is not known. Tocopherol cannot be the only controlling factor here, because some cruciforms, and all campanulates in A. intermedia, are amictic (Gilbert, 1973).

These tocopherol responses do not occur in A. girodi (Gilbert and Litton, 1978) and probably are limited to three or four species in the genus Asplanchna. There-fore, they are not general physiological or developmental effects of the tocopherol molecule. Rather, it appears that it became adaptive in some species of Asplanchna for normal growth processes to be stimulated by tocopherol and for meiotic maturation divisions to be cued to such growth stimulation (Gilbert, in press).

The tocopherol-induced growth just discussed may be greatly exaggerated, and also modified, when certain types of organisms larger than Paramecium are used as prey. The interaction between tocopherol and food type in controlling female size and shape is considered in the next two sections.

Effect of Food Type on Transformations between Tocopherol-dependent Morphotypes in Asplanchna

In both A. intermedia and A. sieboldi, two different female morphotypes may be expressed when dietary tocopherol is available. The cruciform morphotype with its body-wall outgrowths, discussed in the previous section, is the first tocopherol-dependent morphotype to be produced by saccate females (Gilbert, unpublished). In the presence of tocopherol, amictic cruciform females may produce either cruciform- or campanulate-female offspring. The campanulate morphotype is generally somewhat larger than the cruciform morphotype, attaining lengths of 1,200 and 2,000 μm in A. intermedia and A. sieboldi, respectively, and is characterized by a relatively broad corona or head and by the absence of body-wall outgrowths (Fig. 1). The factor which determines whether cruciform or campanulate females will be produced is the type of food eaten.

This food-type effect is most clearly demonstrable in A. intermedia, which can be made to produce exclusively cruciform or campanulate individuals. An experiment from Gilbert (1977a) is summarized below.

Populations of A. intermedia were cultured with $10^{-7}M$ dl-α-tocopheryl phosphate on Brachionus calyciflorus and on Asplanchna brightwelli. Neither food type contained tocopherol, and food was always present in excess. At the end of the experiment, the A. intermedia were preserved. All individuals were morphotyped, and the adults were measured and identified as amictic or mictic. The results are shown in Table 2.

TABLE 2   Dimensions, Morphotypes, and Mode of Reproduction

of Asplanchna intermedia Cultured with $10^{-7}M$ dl-α-tocopheryl

phosphate on two Tocopherol-free Prey--Brachionus calyciflorus

and Asplanchna brightwelli. Number of Individuals Analyzed in

parentheses.

Parameter	Food type	
	Brachionus	Asplanchna
Dimensions (in μm) of adult amictic females (mean ± S.E.)		
1. Body length	897.7 ± 10.5	870.8 ± 7.3
2. Body width	539.7 ± 7.4	506.7 ± 5.2
3. Corona width	446.6 ± 6.9	428.5 ± 4.4
	(80)	(123)
Morphotype of young and adult females	cruciform-campanulate intermediates	campanulates
	(293)	(251)
Proportion of adults mictic	31.6	0
	(117)	(123)

It is clear that the two populations differed markedly from one another in several respects. First, the dimensions of the Brachionus-fed animals were slightly (3-8%) larger than the Asplanchna-fed ones. Second, the Brachionus-fed animals were all cruciform-campanulate intermediates with slight to moderate body-wall outgrowths, while the Asplanchna-fed ones were exclusively campanulate. Third, some of the Brachionus-fed animals were mictic, the ones with the most pronounced outgrowths being mictic, while all of the Asplanchna-fed campanulates were amictic. This difference in reproductive mode is a direct function of the morphotype difference, which, in turn, is directly controlled by the food type.

A comparable effect of diet on morphotype seems to exist in A. sieboldi as well. Campanulates in this species, however, can only rarely be induced and may be either amictic or mictic (Powers, 1912; Gilbert, unpublished). Powers noted that this morphotype appeared in laboratory populations which were cannibalistic or feeding on the water flea Moina paradoxa but not in those which were eating Brachionus. In our laboratory, campanulates of this species are obtained by feeding cruciforms on a smaller species of Asplanchna (A. brightwelli) (Gilbert, unpublished).

The reason why some types of food organisms, such as Asplanchna or Moina, induce campanulates and others, such as Brachionus, do not is completely unknown. Presumably, there is a certain dietary factor in the tissues of the former food types, which, together with tocopherol, is necessary for the induction.

Effect of Food Type on the Size of Tocopherol-dependent Female Morphotypes and Males in Asplanchna

Within a given tocopherol-dependent morphotype, food type can control the extent and proportionality of tocopherol-induced growth, and there is often a correlation between prey size and adult body size and shape.

In A. brightwelli, the size of females containing tocopherol can vary greatly with food type (Gilbert, 1975). In one set of experiments, two populations were cultured with 10^{-7}M d-α-tocopherol, one being fed Paramecium aurelia and the other conspecific females. After 4-5 days on these conditions, 33 adult amictic females from each population were preserved and measured. Means ± S.E. of body length, body width, and corona width were 672 ± 9, 421 ± 7, and 278 ± 8 µm for the Paramecium-fed animals and 750 ± 11, 513 ± 8, and 354 ± 8 µm for the cannibals. Thus, the cannibals were 12, 22, and 27% larger in these dimensions, respectively. The disproportionate increase in corona width is probably an adaptation to facilitate ingestion of large prey.

In A. intermedia, both the size and the biomass of campanulate females varies with the type of congeneric prey eaten (Gilbert, 1975, 1976). In one experiment. means ± S.E. of body length, body width, and corona width were 825 ± 12, 532 ± 9, and 393 ± 8 µm for preserved, adult campanulates fed on A. brightwelli and 942 ± 18, 622 ± 9, and 510 ± 10 µm for those fed on larger, conspecific saccates (Gilbert, 1975). The cannibal campanulates, therefore, were 14, 17, and 42% larger in these dimensions, respectively, than the A. brightwelli-fed ones. The disproportionate increase in the width of the corona, just as in the A. brightwelli cannibals mentioned above, is adaptive in permitting more efficient capture of the larger prey. Other experiments showed that adult campanulates from populations fed on conspecific saccates were considerably heavier than those from populations fed on A. brightwelli, the mean dry weights per female being 7.03 µg (7.06 and 7.00 µg in two replicates from one experiment) and 6.37 µg (6.26, 6.29, 6.41, and 6.57 µg per female in four replicates from another experiment), respectively (Gilbert, 1976).

All cultures in the above-mentioned experiments contained 10^{-7}M d-α-tocopherol and an excess of Asplanchna-prey.

In both A. intermedia and A. sieboldi,the sizes of cruciform females and males is a function of food type (Gilbert, 1975, 1977c). In one experiment (Gilbert, 1977c), populations of both species were cultured in each of two conditions -- Paramecium aurelia with 10^{-7}M d-α-tocopherol, and Brachionus calyciflorus with its tocopherol-containing food source, the alga Euglena gracilis. Adult, mictic, cruciform females and males were preserved and measured. Males are structurally degenerate with only a vestigial digestive system and do not feed. They are completely dependent upon nutrients provided by their mothers and exhibit very little postnatal increase in size (Gilbert, unpublished). Therefore, males were selected at random without regard to age. The results are presented in Table 3.

In both species, the Brachionus-fed females were considerably (11-38%) longer than the Paramecium-fed ones, and the males produced by the former were 17-27% longer than those produced by the latter. Clearly, Paramecium is inferior to Brachionus as a diet, just as was observed for saccate females of A. sieboldi (see above).

TABLE 3   Body Lengths of Males and Cruciform Mictic Females
in Populations of Asplanchna intermedia and A. sieboldi
Cultured on each of two Food Types -- Paramecium aurelia
with 10^{-7}M d-α-tocopherol and Brachionus calyciflorus Fed
on the Tocopherol-containing Alga Euglena gracilis.

Species	Food-type	Sex	Number individuals measured	Mean body length ± S.E. (μm)
intermedia	Paramecium	mictic	35	579 ± 10
	Brachionus	female	30	800 ± 15
sieboldi	Paramecium		34	902 ± 27
	Brachionus		31	999 ± 23
intermedia	Paramecium	male	28	418 ± 9
	Brachionus		33	531 ± 10
sieboldi	Paramecium		46	643 ± 15
	Brachionus		46	751 ± 22

The results of all of the studies discussed in this section show that the body
sizes, weights, and shapes of females of tocopherol-dependent morphotypes are
quite plastic and may be greatly affected by the type of food that the individuals
in the population have been eating. Perhaps the most curious aspect of these
studies is that larger prey often induce larger body sizes and disproportionately
larger coronae. Although it is easy to appreciate that it is adaptive for these
predators to adjust their body size and dimensions in order to most efficiently
capture and ingest available prey, it is extremely difficult to imagine a mechanism by which such accommodation could occur.   The growth promoting effect of
large prey may be due to the chemical nature of their tissues or, to some extent,
to their greater physical size. It should be pointed out, though, that there is
not always a direct relationship between prey size and body size. For example, in
A. sieboldi, cruciforms fed on A. brightwelli and on B. calyciflorus were very
similar in size, and in A. intermedia, cruciform-campanulate intermediates fed on
B. calyciflorus were larger than campanulates fed on the larger A. brightwelli
(Gilbert, 1975) (Table 2).

Although most studies on the effects of food type and hence food size on the body
size of Asplanchna have been performed on tocopherol-dependent morphotypes, some
data have shown that cannibalism in A. intermedia saccates may lead to the production of relatively large bodied forms (Gilbert, 1973). For example, in two
Paramecium-fed populations the mean body lengths ± S.E. were 529 ± 5 and 576 ± 5
μm, while in two populations of cannibals comparable measurements were 630 ± 8 and
657 ± 14 μm, with some individuals attaining lengths of 900 μm. Therefore, accommodation of body size to prey size can also occur in the absence of tocopherol.

Similar, direct relationships between prey size and body size have been reported
in a number of ciliates. Giese (1938), for example, cultured Blepharisma undulans
on a variety of ciliate prey and found that small-sized ones (Tetrahymena and
Khawkinea) caused smaller body sizes than large-sized ones (Stylonichia, Colpidium,
and other Blepharisma).   However, as in Asplanchna, the relationship between prey
size and body size was not perfect; some relatively small prey induced larger body

sizes than some relatively large ones.

Effect of Food Type on the Morphology of Brachionus calyciflorus

B. calyciflorus has a lorica or skeleton which may possess posterolateral spines
of variable length. The study described here was conducted to determine if the
qualitative nature of the diet could influence lorica length or spine length.

Populations were cultured xenically for many generations on each of three food
organisms -- the bacterium Aerobacter aerogenes, the yeast Rhodotorula glutinis,
and the alga Euglena gracilis. These diets were each provided at a level of 100
μg dry weight per ml. The populations were maintained in constant light as
described elsewhere (Gilbert, 1975), were changed to fresh suspensions of food
each day, and were not permitted to exceed a density of about 7 animals per ml.
On one sampling date, rotifers from three replicate populations on each food type
were preserved in 10% formalin. The length of the lorica and the mean length of
the two posterolateral spines were measured at 200 magnifications to the nearest
6 μm. The data are presented in Table 4.

TABLE 4   Lorica Lengths and Ratios of Spine Length to Lorica
Length of Adult, Amictic-female Brachionus calyciflorus from
Populations Cultured with 100 μg ml^{-1} of the Yeast Rhodotorula
glutinis, the Bacterium Aerobacter aerogenes, and the Alga
Euglena gracilis at 23°C.

Food	Repli-cate	Number measured	Lorica length in μm (mean ± S.E.)	Spine length: lorica length ratio (mean ± S.E.)
Rhodotorula	1	28	169.5 ± 2.5	0.134 ± .006
	2	35	174.7 ± 2.5	0.159 ± .010
	3	35	172.9 ± 1.4	0.147 ± .008
Aerobacter	1	36	179.2 ± 2.3	0.108 ± .003
	2	37	173.3 ± 2.1	0.111 ± .005
	3	22	183.0 ± 2.5	0.091 ± .004
Euglena	1	32	182.2 ± 2.7	0.139 ± .005
	2	21	178.8 ± 3.1	0.137 ± .005
	3	25	181.6 ± 2.0	0.147 ± .007

It is clear that food type had little effect on body length, although the
Rhodotorula-fed individuals generally were somewhat (4-5%) smaller than Aerobacter-
or Euglena-fed ones. Food type, however, had a marked effect on the length of the
posterolateral spines relative to that of the body. Females cultured on
Rhodotorula and on Euglena had spine length to body length ratios 37% and 43%
higher than those cultured on Aerobacter, respectively.

The basis of the relationship between food type and spine length in B. calyciflorus
is unknown. Spine length has been shown to vary inversely with food level
(Schneider, 1937; Buchner, Mulzer, and Rauh, 1957; Erman, 1962; Rauh, 1963;
Halbach, 1970). Therefore, even though the three food types were available in
equivalent amounts, the Aerobacter may have been eaten or assimilated with greater

efficiency than the other food organisms, causing the production of shorter spines.

The mechanism by which diet may influence spine length is not understood. Spines are evaginations of the body wall, just as the body-wall outgrowths of Asplanchna, and develop through localized hypodermal growth. It is not known if such growth involves additional nuclear divisions in the hypodermis.

Effect of Food Type on the Pattern of Resting Egg Hatching in Brachionus calyciflorus

If the haploid egg of a mictic female is fertilized, it undergoes partial embryological development, secretes a complex multilayered shell, and then becomes dormant. The so-called resting egg, then, is really an encysted embryo in an arrested stage of development. When dormancy is broken, the embryo resumes development and hatches into an amictic female. The period of dormancy is extremely variable. The biology of resting eggs has been reviewed elsewhere (Gilbert, 1974b), and their ultrastructure has recently been described (Wurdak, Gilbert, and Jagels, 1978).

The purpose of the experiment described here was to determine if the diet of fertilized mictic females could affect the hatchability or hatching pattern of the resting eggs they produced. Since resting eggs require much more energy to produce than diploid parthenogenetic eggs (Gilbert, in press), the nature of the diet may have an especially pronounced effect on their development and viability.

B. calyciflorus was cultured in constant light on each of two food types -- the yeast Rhodotorula and the alga Euglena gracilis -- as described elsewhere (Gilbert, 1975). Mictic females were induced by maintaining the cultures at high population densities (Gilbert, 1963), and mature, dormant resting eggs, 0-2 days of age, were removed from these cultures, washed, and then stored in inorganic rotifer medium for 1 day at 3°C in the dark. Each of three replicate batches of 24-26 eggs produced by the Brachionus on the two diets was then placed in a 35 X 10 mm plastic Petri dish containing 5 ml rotifer medium kept at 22°C under constant illumination. Each day, the dishes were examined, and every animal that hatched was removed and recorded. The hatching patterns observed are shown in Table 5.

At the end of 15 days, almost all of the eggs had hatched, but more of the eggs from Euglena-fed females hatched than those from Rhodotorula-fed ones. The mean values ± S.E. for percent hatchings were 96 ± 2 and 83 ± 5% respectively, and are clearly different from one another. Row-by-column tests of independence showed that the proportions of eggs hatching were not significantly different among the three replicates of each set of eggs ($p > 0.05$) but that the pooled values for the two sets were significantly different from each other ($G = 7.49$ with 1 df; $p = 6.2$ X 10^{-3}).

The pattern of hatching for the resting eggs from the females of the two diets was also very different. Row-by-column tests of independence on the numbers of hatchings in successive days showed that there were no significant differences among the replicates for each set of eggs ($p > 0.05$) but that the pooled values for the sets were very different from each other ($G = 62.13$ with 8 df; $p < 1$ X 10^{-9}).

Inspection of the patterns shows that the eggs from Euglena-fed females hatched more quickly and over a greater span of time than those from Rhodotorula-fed females.

These results demonstrate that the maternal diet may influence the hatchability

TABLE 5 Hatchability and Hatching Pattern of Resting Eggs of Brachionus calyciflorus Cultured on the Alga Euglena gracilis and on the Yeast Rhodotorula glutinis.

Parental diet	Replicate	Number eggs at start	Number eggs hatching on successive days															Percent hatched
			1	2	3	4	5	6	7	8	9	10	11	12	13	14	15	
Euglena	1	25	0	0	6	1	1	1	2	1	5	1	2	1	0	2	0	92
	2	24	0	0	8	5	2	0	1	2	2	0	1	0	3	0	1	100
	3	26	0	0	12	2	1	1	0	0	1	2	0	3	2	0	1	96
	1-3	75	0	0	26	8	4	2	3	3	8	3	3	4	5	2	1	96
Rhodotorula	1	26	0	0	4	1	4	5	5	4	0	0	0	0	0	0	0	88
	2	24	0	0	2	1	3	7	5	2	0	0	1	0	0	0	0	88
	3	25	0	0	3	1	2	6	4	1	0	0	0	0	0	0	1	72
	1-3	75	0	0	9	3	9	18	14	7	0	0	1	0	0	0	1	83

and the hatching pattern of resting eggs. The manner in which diet may alter hatching is not known, but it is likely that the quality and quantity of yolk and other reserves produced by the vitellarium and transported to the fertilized oocyte may affect both the viability of the resting egg embryo and its response to those environmental factors controlling the breaking of dormancy.

CONCLUSIONS

The results presented in this paper show that the kind of food eaten by Asplanchna and Brachionus may significantly affect their size, shape, population dynamics, mode of reproduction, and emergence from dormancy.

Food type can influence the biology of Asplanchna in many different ways. First, the tocopherol content of the food plays an important role in controlling the morphotype and reproductive mode of the female. If the food contains no tocopherol, females will be relatively small and saccate and will produce only female offspring. If the food contains enough tocopherol for above-threshold levels of this compound to be assimilated, females will be much larger, may have body-wall outgrowths, and may produce male offspring. Second, in A. intermedia and A. sieboldi, food type determines which of the two possible tocopherol-dependent, female morphotypes will be expressed. Campanulate females have only been produced on diets of congeneric or crustacean prey. Third, the size and shape of females of a given morphotype are a function of the food type. Larger prey typically induce larger females with disproportionately wider heads. Fourth, the type of food eaten by mictic females may influence the size of their male offspring; Brachionus-fed females, for example, produce larger males than Paramecium-fed ones. Fifth, food type may dramatically affect the fecundity, longevity, and reproductive rate of females. In A. sieboldi, saccate females fed on Brachionus produced more than twice as many offspring, lived longer, and reproduced much more rapidly than those fed on Paramecium.

Some information is available on the mechanisms by which tocopherol influences the development of Asplanchna. This molecule increases body size and alters body shape by inducing additional mitotic divisions, cytoplasmic growth, and nuclear DNA endoreduplication. Haploid males are produced parthenogenetically following meiotic maturation divisions, but the factors which induce such divisions are not known. Tocopherol must be present for meiosis to occur, but meiosis does not occur in all tocopherol-containing females.

The mechanisms responsible for the other food-type effects noted in Asplanchna are not understood. Presumably, these effects are due to specific, qualitative characteristics of the different foods, but, in some cases, they may be at least partially explained by the amounts of material ingested or assimilated when different food organisms are available. For example, Brachionus may be more efficiently captured and assimilated than Paramecium and, therefore, may be a better diet. Also, the intriguing fact that cannibalistic A. intermedia campanulates are larger when they eat larger congeneric prey may be due to the physical size rather than the nutritional quality of the prey.

Food type can influence the biology of Brachionus calyciflorus in several ways. First, the length of the body and especially the posterolateral spines may be a function of food type. For example, animals fed on Aerobacter had much shorter spines than those fed on either Rhodotorula or Euglena, even though the biomass of food available in each case was the same. Second, the ability of resting eggs to hatch and the time required for such eggs to hatch may vary with the type of food eaten by the females producing these eggs. The bases for these food-type effects are unknown. Presumably, the three cell types are qualitatively different in nutritional content, but they may also differ in the efficiencies at which they are ingested and assimilated.

REFERENCES

Beauchamp, P. de (1909). Recherches sur les rotifères: les formations tégumentaires et l'appareil digestif. Arch. Zool. Exper. Gen., 10, 1-410.

Birky, C. W., Jr., and J. J. Gilbert (1972). Vitamin E as an extrinsic and intrinsic signal controlling development in the rotifer Asplanchna: uptake, transmission, and localization of H^3-α-tocopherol. J. Embryol. Exp. Morph., 27, 103-120.

Buchner, H., Mulzer, F., and F. Rauh (1957). Untersuchungen über die Variabilität der Rädertiere. I. Problemstellung und vorläufige Mitteilung über die Ergebnisse. Biol. Zbl., 76, 289-315.

Draper, H. H. (1970). The tocopherols. In R. A. Morton (Ed.), Fat-soluble Vitamins, Vol. 9. Pergamon Press, Oxford, pp. 333-393.

Erman, L. A. (1962). Cyclomorphosis and feeding of plankton Rotifera (in Russian). Zool. Zh., 41, 998-1003.

Giese, A. C. (1938). Cannibalism and gigantism in Blepharisma. Trans. Amer. Micros. Soc., 57, 245-255.

Gilbert, J. J. (1963). Mictic female production in the rotifer Brachionus calyciflorus. J. Exp. Zool., 153, 113-124.

Gilbert, J. J. (1973). The induction and ecological significance of gigantism in the rotifer Asplanchna sieboldi. Science, 181, 63-66.

Gilbert, J. J. (1974a). Effect of tocopherol on the growth and development of rotifers. Amer. J. Clin. Nutr., 27, 1005-1016.

Gilbert, J. J. (1974b). Dormancy in rotifers. Trans. Amer. Micros. Soc., 93, 490-513.

Gilbert, J. J. (1975). Polymorphism and sexuality in the rotifer Asplanchna, with special reference to the effects of prey-type and clonal variation. Arch. Hydrobiol., 75, 442-483.

Gilbert, J. J. (1976). Polymorphism in the rotifer Asplanchna sieboldi: biomass, growth and reproductive rate of the saccate and campanulate morphotypes. Ecology, 57, 542-551.

Gilbert, J.J. (1977a). Effect of the non-tocopherol component of the diet on polymorphism, sexuality, biomass, and reproductive rate of the rotifer Asplanchna sieboldi. Arch. Hydrobiol., 80, 375-397.

Gilbert, J. J. (1977b). Mictic-female production in monogonont rotifers. Arch. Hydrobiol. Beih., 8, 142-155.

Gilbert, J. J. (1977c). Defenses of males against cannibalism in the rotifer Asplanchna: size, shape, and failure to elicit tactile feeding responses. Ecology, 58, 1128-1135.

Gilbert, J. J. Female polymorphism and sexual reproduction in the rotifer Asplanchna: evolution of their relationship and control by dietary tocopherol. Amer. Nat., in press.

Gilbert, J. J. and J. R. Litton, Jr. (1978). Sexual reproduction in the rotifer Asplanchna girodi: effects of tocopherol and population density. J. Exp. Zool., 204, 113-122.

Gilbert, J. J. and G. A. Thompson, Jr. (1968). Alpha-tocopherol control of sexuality and polymorphism in the rotifer Asplanchna. Science, 159, 734-736.

Gilbert, J. J. and E. S. Wurdak (1978). Species-specific morphology of resting eggs in the rotifer Asplanchna. Trans. Amer. Micros. Soc., 97, 330-339.

Gilbert, J. J., Birky, C.W., Jr. and E. S. Wurdak. Taxonomic relationships of Asplanchna brightwelli, A. intermedia, and A. sieboldi. Arch. Hydrobiol., in press.

Halbach, U. (1970). Die Ursachen der Temporalvariation von Brachionus calyciflorus Pallas (Rotatoria). Oecologia, 4, 262-318.

Jones, P. A. and J. J. Gilbert (1977). Polymorphism and polyploidy in the rotifer Asplanchna sieboldi: relative nuclear DNA contents in tissues of saccate and campanulate females. J. Exp. Zool., 201, 163-168.

Kabay, M. E., and J. J. Gilbert (1977). Polymorphism and reproductive mode in the rotifer, Asplanchna sieboldi: relationship between meiotic oogenesis and shape of body-wall outgrowths. J. Exp. Zool., 201, 21-28.

King, C. E. (1967). Food, age, and the dynamics of a laboratory population of rotifers. Ecology, 48, 111-128.

Litton, J. R., Jr., and J. J. Gilbert (1975). Analysis of tocopherol in Rhodotorula glutinis, Agaricus campestris, and Euglena gracilis using spectro-fluorometry and rotifer bioassay. J. Gen. Appl. Microbiol., 21, 345-354.

Powers, J. H. (1912). A case of polymorphism in Asplanchna simulating a mutation. Amer. Nat., 46, 441-462, 526-552.

Rauh, F. (1963). Untersuchungen über die Variabilität der Rädertiere. III. Die experimentelle Beeinflussung der Variation von Brachionus calyciflorus und Brachionus capsuliflorus. Z. Morph. Ökol. Tiere, 53, 61-106.

Schneider, P. (1937). Sur la variabilité de Brachionus pala Ehrenberg dans les conditions expérimentales. C. r. Soc. Biol., 125, 450-452.

Steinböck, O. (1958). Zur Phylogenie der Gastrotrichen. Verh. Deut. Zool. Ges. Graz,1957, 128-169.

Theilacker, G. H. and M. F. McMaster (1971) Mass culture of the rotifer Brachionus plicatilis and its evaluation as a food for larval anchovies. Mar. Biol., 10, 183-188.

Wurdak, E. S. and J. J. Gilbert (1976). Polymorphism in the rotifer Asplanchna sieboldi. Fine structure of saccate, cruciform and campanulate females. Cell Tiss. Res., 169, 435-448.

Wurdak, E. S., J. J. Gilbert and R. Jagels (1978). Fine structure of the resting eggs of the rotifers Brachionus calyciflorus and Asplanchna sieboldi. Trans. Amer. Micros. Soc., 97, 49-72.

POPULATION DYNAMICS OF CTENOPHORES IN LARGE SCALE ENCLOSURES OVER SEVERAL YEARS

M. R. REEVE

Rosenstiel School of Marine and Atmospheric Science, University of
Miami, 4600 Rickenbacker Causeway, Miami, Florida 33149, U.S.A.

ABSTRACT

For five years (1974-78) populations of the ctenophores <u>Pleurobrachia</u> and <u>Bolinopsis</u>
have occurred in large (70-1,300 m^3) transparent containers along with other
components of the natural pelagic ecosystem of Saanich Inlet, British Columbia,
which was the site of the study. This report reviews the population dynamics of the
ctenophores, particularly in relation to growth, fecundity and mortality, as
affected by different experimental conditions, such as numbers and kinds of prey
organisms, and discusses how interactions between ctenophores and their prey can
produce conditions which both promote and inhibit the successful development of the
carnivore populations.

KEYWORDS

Ctenophores; <u>Pleurobrachia</u>; <u>Bolinopsis</u>; CEPEX experiments; population dynamics;
zooplankton feeding; predator-prey interactions.

INTRODUCTION

The CEPEX (Controlled Ecosystem Population Experiment) program was a field study of
water column ecosystem dynamics as modified by natural and artificial stresses,
which extended over five summers (1974-78) in Saanich Inlet, British Columbia.
Natural marine plankton populations were captured by raising large transparent
polyethylene bags upwards through the water column to the surface. Originally
(1974-75) up to six bags or CEEs (Controlled Experimental Ecosystems) were deployed
in each experiment, having a volume of 68 m^3, a diameter of 2.5 m and length of 14 m.
Subsequently, much larger CEEs were used (1,300 m^3) which were 10 m in diameter and
23.5 m in length. The objectives of the program, details of construction of CEEs
and description of experiments are to be found in numerous publications, including
Reeve and others (1976), Grice and others (1977) and Menzel and Steele (1978).

During these experiments populations of ctenophores often developed, sometimes
having profound effects on the rest of the ecosystem. As time progressed, we also
progressed in the degree of sophistication of our collection, handling and
experimental techniques for these delicate animals, and consequently, our
understanding of their population dynamics, and interactions with other components
of the ecosystem. Although some of the experiments involved the addition of toxic
substances during the first four years, there was always a control CEE for any

73

experiment to which no such additions were made. Except where specifically noted, all information below refers to populations of unpolluted CEEs.

STATUS OF NUTRITIONAL RESEARCH IN CTENOPHORES

Reeve and Walter (1978) reviewed recent research on the nutritional ecology of ctenophores, noting that most quantitative information beyond distributional data and studies on feeding mechanisms were largely limited to the past 15 years. Laboratory studies showed that ctenophores were unusual in their potentially extremely high growth and fecundity, although such rates were only seen at food levels much higher than normally associated with the natural environment. Unlike many zooplankton, they did not become satiated at moderate food levels but could continue to increase their food consumption even beyond a daily ration of 1,000% of their bodily carbon content. Their efficiency of conversion of food into their own body tissues, however, was less than 10% because of their high energetic requirements. Although usually efficient in food digestion, at extremely high food concentrations their assimilation efficiency was progressively reduced.

Very little is known, however, about how such experimental information relates to interpretation of fluctuations of environmental ctenophore populations. At modest environmental prey concentrations such as those which usually occurred in CEEs, animals in the laboratory tended to grow very slowly, or even shrink.   Their population outbursts seemed to require far higher prey concentrations than were ever seen in the environment. Detailed information on environmental population dynamics is very limited. Often they were regarded as nuisances, to be removed from plankton samples or excluded if possible, and when estimated it was usually on the basis of total volume, or total numbers, regardless of size.   The common lobate genera Bolinopsis and Mnemiopsis quickly dissolve in formalin. Bolinopsis is particularly susceptible to damage by net collection and almost invariably killed or mutilated. Reeve and Baker (1975) described the use of a gentle sieving technique on board the collecting boat to obtain abundance by volume of different size classes, but even this severely underesti.nated the numbers of smallest animals. Pleurobrachia, a common tentaculate genus, generally preserves well, except for the smallest larvae which do not preserve quantitatively. Added to these difficulties of environmental assessment is the fact that the large relative size of ctenophores means that they are numerically rare compared to their prey organisms, and like them, not uniformly distributed.   Patchiness magnifies the difficulty of obtaining a representative sample of the population.   It has been impossible, therefore, to extrapolate laboratory experimental data to the environment to explain quantitatively predator/prey relationships, partly because of the uncertainty of the significance of laboratory measurements to nature, and partly because of the uncertainty of the population structure and dynamics of the ctenophores.

We are left at present with numerous historical records relating ctenophore abundance with virtual absence of herbivores (see Reeve and Walter, 1978) and more recent model projections (e.g. Kremer, 1979; Reeve, Walter and Ikeda, 1978) suggesting that indeed, ctenophores could be capable of the decimation of herbivores under certain conditions. As Kremer (1979) and Reeve and Walter (1976) pointed out, however, there are often fluctuations in co-occurring ctenophore and copepod populations which cannot be readily explained.

THE UTILITY OF LARGE ENCLOSURES

Large enclosures provide a unique opportunity for the study of carnivorous zooplankton. In the first place the huge problem of advection and patchiness becomes much more manageable, since week after week the populations are confined to the same bounded volume of water, which can be revisited at will. Whatever other problems of population sampling and analysis remain, at least there is the certainty that the

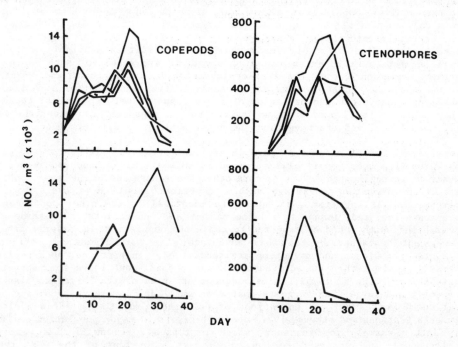

Fig. 1.   (Upper) Changes in copepod and ctenophore munbers in
four similarly treated CEEs.   (Lower) Changes in two
dissimilarly treated CEEs (see text).

Fig. 2.   (Left) Large cod-end reservoir attached to net.
(Right) Reservoir and outer bag partly removed.

same population is being sampled. The facilities required to maintain and provide food for carnivore populations, individuals of which can exceed 100 mm in length, is beyond the scope of all but the most specialized laboratories. The CEPEX enclosures are essentially automatic culture systems, not only for the carnivore populations themselves but also for their food. With such facilities available, only modest facilities for specific experimental observations are required, such as the small constant temperature room and trailer-laboratories available at the CEPEX site. Although the CEEs themselves are not amenable to the close control of a laboratory experiment, conditions can be manipulated and compared between several containers. With a depth up to 23.5 m, shallow water plankton organisms can undertake many of their normal diel and other vertical behavior patterns and interactions.

In any discussion of results of experiments, whether in the laboratory or in large enclosures, the question is always raised as to what extent observed events in one container are duplicated in one or more other containers, under the same conditions. This is the important advantage such a container should have over the natural environment. At the start, it should be identical to the other containers, and having been isolated, immune to all the random (i.e. unpredictable) outside sources of variation which could be expected to cause its populations to diverge from those of adjacent containers. For this advantage, the naturalness of vertical and horizontal turbulence and advection are sacrificed. In practice, such replication of events in similarly managed CEEs did occur in small containers for microplankton (Takahashi and others, 1975), and subsequent experience with larger zooplankton in both small and large CEEs suggests that even small deliberate changes, such as the addition of very small concentrations of pollutants, or the variable addition of nutrients (Parsons and others, 1977) does not result in major population shifts. In this latter example for instance, which is typical, six small CEEs were raised simultaneously and observed for up to 40 days. In four of the CEEs the only differences between them was that nutrient additions were made (Parsons and others, 1977), ranging from 0 to 10 µg-A N. A large initial nutrient level in the water and nutrient recycling at a rapid rate in all the CEEs was presumed to account for the fact that changes in both copepod and ctenophore numbers were very similar over the period (Fig. 1 upper). In the other two CEEs an attempt was made (Reeve and Walter, 1976) to remove developing ctenophore populations by net collection from one and transfer them to the other. This perturbation quickly produced very drastic changes in both copepod and ctenophore populations (Fig. 1 lower). The behavior of larger zooplankton in containers tends to support the hypothesis that population changes can be ascribed to effects other than random noise.

THE CTENOPHORES

The ctenophores of Saanich Inlet are represented by the genera Pleurobrachia and Bolinopsis. Only very rarely were members of the genus Beroe encountered, but these never occurred in the CEEs. There is some uncertainty regarding the specific status of many ctenophores, partly due to the problems of comparison of specimens which cannot be preserved, and sometimes the same specific name is used for animals both from the Atlantic and Pacific. Following other workers, Pleurobrachia is tentatively assigned as P. bachei Agassiz (see Hirota, 1974) and Bolinopsis as B. infundibulum (Martens). Pleurobrachia is a member of the order Cydippida, which feeds by trapping prey organisms by means of a pair of long tentacles which may be extended many times the length of its globular body, which are then retracted to its mouth. Bolinopsis of the order Lobata, although it begins life as a tentaculate larva, very soon begins to develop lobes which project below its mouth, upon the inner sides of which prey organisms impinge, and are apparently restrained by mucus and transferred to the mouth. Their method of feeding is described in more detail in Reeve and Walter (1978). Both species are hermaphrodite, and capable of self fertilization, the newly-hatched larvae beginning life at only 0.3 mm in diameter. The two species are distinguishable even at this stage and by 1-2 mm, Bolinopsis shows the first signs of lobe development. Its tentacles gradually atrophy and it becomes fully lobate at a length of 10-15 mm (including lobes). The maximum diameter (oral/aboral) attained by Pleurobrachia in CEEs was 15 mm, and the maximum length (including lobes) reached by Bolinopsis was about 140 mm.

COLLECTION METHODS

Throughout the five year period from 1974-78, one standard collection technique persisted. This was a vertical haul up the center of CEEs using a Bongo net pair of 200 μm mesh and either 20 or 40 cm mouth diameter, depending on whether sampling in small or large CEEs. The sample was preserved in 2% buffered formalin. During the 1974 field season, it became clear that Bolinopsis was present in the CEEs but because it failed to preserve, no record of it remained when the preserved samples were analyzed. From 1975 onwards, live plankton samples were returned to the laboratory. In that year (in the small CEEs) they were collected using a 500-μm mesh net of 50-cm mouth diameter. The length of each indivdual was measured from its mouth to the top of its body. Lobes were nearly always either damaged or missing and so were not included. A length/dry weight/organic carbon relationship was developed in the laboratory so that biomass of populations collected could be calculated. From 1976 onwards, when only larger CEEs were sampled, a considerably larger net (1-m mouth diameter) was used with a much larger cod-end. This cod-end (Fig. 2) consisted of a cylindrical transparent acrylic reservoir of 30-liter capacity protected by an outer mesh bag, which also served as the means of attachment to the net by means of a zipper. Once animals were captured in the cod-end reservoir, they remained there until arrival back at the laboratory. The net was slowly lowered to the bottom of the CEE, allowing the mouth to completely collapse down, to rest near the bottom of the cone of the CEE. This proved necessary because both previous SCUBA observation and net collections showed that large Bolinopsis tended to congregate very close to the bottom of the CEE. After a few minutes the net was very slowly hauled back to the surface at a speed of 5 m min^{-1}. When large populations were present, the reservoir was returned to the laboratory immediately because population analysis was time consuming and deterioration of the animals could be rapid. If populations were sparse, all three CEEs could be sampled sequentially (using different reservoirs) and then returned for analysis. Since the sampled volume was large (20 m^3) compared to the Bongo tows (5 m^3), the larger (and rarer) Pleurobrachia were also measured, their numbers to be integrated with those from the preserved samples. Bolinopsis less than 5 mm in length were examined by removing 2-liter aliquots from the reservoir.

In 1977, laboratory tests showed that very young (less than 1 mm) Pleurobrachia did not preserve reliably. In 1978, in addition to the other sampling procedures, A 20-liter water sample was obtained using a pump and hose lowered through the water column. On return to the laboratory, the sample was slowly and gently concentrated, and all the ctenophores in it measured. This technique produced an order of magnitude more Pleurobrachia larvae less than 2 mm than have been previously encountered using preserved material. Very few Bolinopsis larvae, on the other hand, were seen. It is probable that the pump destroyed them.

Although the quantitative collection and analysis of ctenophore populations, therefore, has reached a level more advanced than performed by any workers in the natural environment, these difficult animals still are not being collected completely reliably.

BEHAVIOR OF CTENOPHORE POPULATIONS IN LARGE ENCLOSURES

Ctenophore/Copepod Interactions

Greve (1972) reported experiments in small dishes in which the large copepod Calanus caused mortality of larval ctenophores, and subsequently showed such effects in containers of about 1 liter (Greve, 1977), although no details of densities of organisms were provided. Reeve and Walter (1978) referred to an adverse affect of the much smaller copepod Acartia tonsa on larval ctenophores at very high densities (100,000 m^{-3} in large (30-liter) containers)) and Stanlaw, Reeve and Walter (in press) summarized the results of extensive series of experiments in volumes no less than 4 liters which suggested that adults, and even copepodites of Acartia tonsa, could cause mortality of ctenophore larvae up to 1.5 mm, but densities of adults needed to be over 20,000 m^{-3} and copepodites much higher. Larger copepods such as Undinula or Calanus, however, could achieve this effect at much lower densities (1,000 m^{-3}).

Figure 3 provides a summary of CEPEX experiments in which there were initial captures of Pleurobrachia. From left to right, the populations graphed are "small copepods" (mostly Acartia, Pseudocalanus and Paracalanus in the 200-μm net) Calanus, larval ctenophores (Pleurobrachia less than 2 mm preserved in the 200-μm net) and "post-larvae" (all other Pleurobrachia). The top three distributions cover experiments lasting up to 40 days in small CEEs in 1974 (A and B) and 1975. The lower three distributions are from large CEE experiments of 1976-78, lasting up to 80 days. In all of these experiments, started in June or later, small populations of Pleurobrachia were captured. Except for the 1974 experiments, the capture of post-larval animals was very low, usually less than 1 m^{-3}, or less than 0.1 m^{-3} of animals of reproductive age (8 mm or more).

The experiments follow two distinct patterns. The first type of pattern occurs in 1974 A and B, 1975 and 1978. Calanus populations were very low (less than 1,000 m^{-3}) and small copepods quickly reached a peak between 10,000-20,000 m^{-3}. Ctenophore larvae quickly rose and then fell again and post larvae followed the same pattern. A more detailed analysis of events for a typical year (1978) showed that the initial capture of ctenophores was too low to prevent rapid reproduction and population build-up of the small copepods. Within a few days, however, large numbers of ctenophore larvae had been produced which grew quickly at first in response to the increased food level. By this time the copepods were being grazed drastically, their numbers declining rapidly to a point which restricted the further reproduction of ctenophores, and particularly in the longer experiment (1978) also reduced the larger ctenophores. In this experiment populations were monitored for a total of 111 days, but enough ctenophores remained to prevent any resurgence of copepod populations. Presumably, in these enclosed containers, the copepods could never become uncoupled from their predators, as might have happened in the patchy natural environment.

The second pattern is exemplified by experiments of 1976 and 1977. In this case Calanus population peaks exceeded 8,000 m^{-3}. It is likely that such numbers were present in the CEEs over most of the period. A series of samples designed to investigate vertical distribution in 1976 suggested that much of the time Calanus remained very near the bottom of the CEEs, where they were not sampled adequately, and only came within daytime sampling range occasionally. Routine sampling was from a depth 20 m, although the bottom of the cone was at 23.5 m. It was also shown that Pleurobrachia larvae tend to congregate near the bottom. Small copepod populations increased dramatically in both experiments to near 40,000 m^{-3} and declined only slowly to 10,000 m^{-3}. The small copepod increase was clearly due to reproduction and the rapid growth of cohorts. In both experiments, although ctenophores were initially present and in evidence throughout the experiments in very low numbers, no significant populations developed. This, despite the highest densities of potential

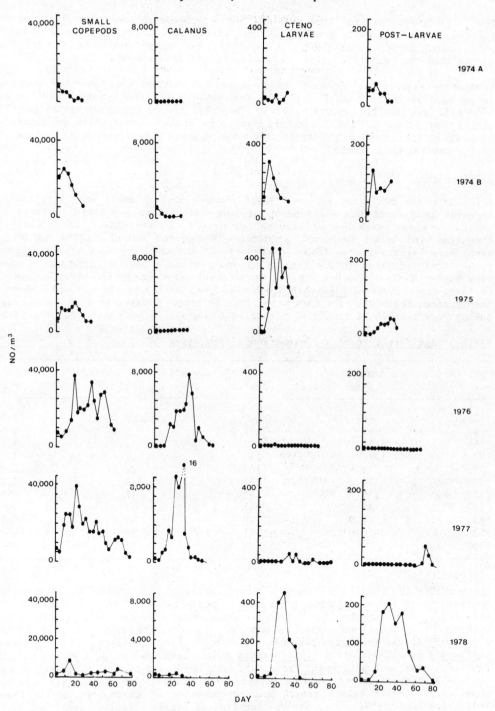

Fig. 3.   Changes in populations of "small" copepods,
Calanus, Pleurobrachia larvae and post larvae
during 6 experiments in enclosures.

food organisms of all the experiments. Figure 4 shows an adult "small" copepod on the same scale as a newly-hatched (0.3 mm) ctenophore larva and a copepod nauplius which could serve as its food. A Calanus-sized copepod would be several times larger than the "small" copepod, depending on its developmental stage.

Laboratory experiments reported above suggested that on the occasions when very large populations of small copepods and Calanus occurred, they could prevent successful reproduction of ctenophores. They effectively achieved this on the CEPEX scale. By virtue of the laboratory and large scale demonstration of control by copepods of their natural predators, the concept gains more credence as a phenomenon in the natural environment.

Other Causes of Mortality

Aside from the two occasions when large numbers of both small and large copepods appeared to prevent the development of large ctenophore populations, there were never any known predators of ctenophores present in the CEEs. Indeed very few organisms are known as usual predators (Reeve and Walter, 1978) except the ctenophore Beroe and some fish. Nevertheless, the age structure of Pleurobrachia populations indicated that mortality of young animals as they grew was routinely very high. Table 1 shows a typical development of a population over 34 days. At capture, the numbers of Pleurobrachia are so low that the sample volume of 20 m^{-3} is hardly adequate to describe them properly. No larvae (less than 2 mm) were present and animals capable of producing offspring (8 mm and larger) were less than 1 m^{-3}.

TABLE 1 Age Structure of a Pleurobrachia Population

Day	Larvae < 2 mm	Juveniles 2-4 mm	4-8 mm	Mature > 8 mm
2	--	0.1	0.4	0.8
9	2697	1.3	0.6	0.9
16	4153	23	5.0	1.1
23	1774	156	26	1.1
30	718	183	21	1.0
37	224	137	14	1.7
43	193	162	18	--
51	32	67	11	--
62	2.9	20	9.2	0.2
68	0.8	21	16	--
76	--	6.0	10	--

These mature individuals never significantly exceeded this level. Within nine days, however, these individuals had succeeded in producing vast numbers of larvae which exceeded 4,000 m^{-3} by day 16. The larvae showed very poor survival, less than 10% of them reaching 2 mm, and probably none of the reaching maturity. By contrast Bolinopsis larvae reached a peak of only 30 m^{-3}. Since the total carbon biomass of Bolinopsis equalled that of Pleurobrachia at its peak (5.5 mg m^{-3}) it may be safely concluded that most Bolinopsis larvae had been destroyed by the collection process. By day 30, both adult and naupliar copepods had become reduced to levels well below those which permit either growth and reproduction of ctenophore adults (Reeve, Walter and Ikeda, 1978) or growth and survival of larvae (Stanlaw, Reeve and Walter, in press) in laboratory experiments. Indeed even peak densities of copepod nauplii and adults over that period were marginal for such growth. This may be misleading to the extent that average copepod densities throughout the water column of the CEEs does not take into account the fact that they were probably in layered or patchy

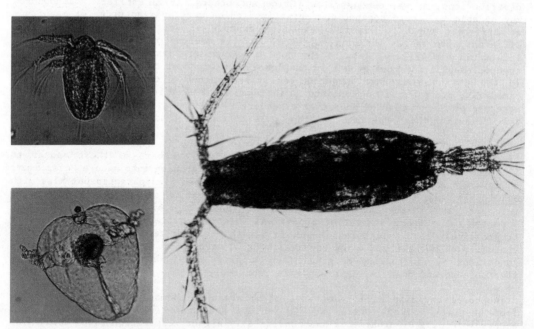

Fig. 4. Relative sizes of typical adult "small" copepod (1 mm), newly-hatched ctenophore larva (0.3 mm), and nauplius capable of being ingested by the larva.

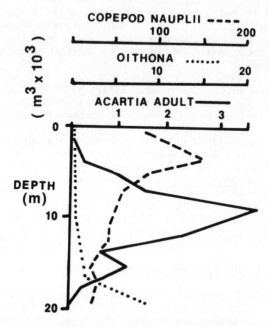

Fig. 5. Layered distributions of three differently-sized potential food organisms of ctenophores to 20 m.

distributions, as was demonstrated (Grice and others, 1977) earlier. Ctenophores may be able to locate copepod concentrations which vary over an order of magnitude at different depths.   Figure 5 shows an example of vertical layering of three different size ranges (nauplii, Oithona, Acartia adults) on one occasion in 1976. This would explain why, despite the need of ctenophore larvae for 500,000 m^{-3} of copepod nauplii for good growth in the laboratory, some of them, at least, survived in the CEE when average nauplius densities reached a maximum of 100,000 m^{-3}. Similarly, ctenophores can continue to increase their food intake and growth rate to make use of 200-μm copepod densities exceeding 100,000 m^{-3}, but average densities in the CEE during this experiment reached only a fifth of this.

Although mortality is striking among the early stages, it appears to be much less significant in the older juveniles (4-8 mm).   Small numbers of these appear to linger on throughout the rest of the period, neither growing into mature (over 8 mm) animals nor dying off. This was the case not only with Pleurobrachia but also with Bolinopsis. Experiments designed to test the starvation capacity of both species were performed in the laboratory during this time (Fig. 6) on 4 mm animals taken from the CEE. Under starvation conditions (membrane-filtered water) both species progressively lost weight and shrank in length, but for more than 10 days showed virtually no mortality. A few survived over 20 days under these conditions. Under conditions of fluctuating marginal food supply, individual ctenophores may well fluctuate up and down in their own size repeatedly.

Ctenophores may also have other ways of maintaining themselves during periods of food shortage.   Ikeda (1977) and Gyllenberg and Greve (1979) both showed that actively feeding Pleurobrachia had considerably higher respiration rates than those not feeding. Lobate ctenophores demonstrated the same ability. Figure 7 shows a typical result of the effect of starving a closely related lobate ctenophore Menemiopsis in the laboratory over several days.   In this case intial respiration measurements of well-fed animals were not performed when they had food in their guts, i.e. they were not actually digesting at the time. In the absence of food, respiration decreased by half, and remained about that level, unless the animals were again provided with food. It appears then that most ctenophores can rapidly adjust their metabolic rates to conserve their organic body tissue in times of food shortage, and reduce their rate of shrinkage and lengthen their survival times under these adverse conditions.

It is generally considered (see Reeve and Walter, 1978) that ctenophores depend on crustacean food.   We observed, however, at times when large concentrations of diatoms and ciliates were present in CEEs that traces of these could be seen inside the guts of Bolinopsis, although not in Pleurobrachia.   Since the lobes of Bolinopsis employ a food collection technique which depends on impingement on the lobes and subsequent mucous entanglement and ingestion, it should be possible for any particle to be ingested. We never observed any more than tiny accumulations of such cells in the guts of Bolinopsis, and laboratory incubation in such water showed they could not maintain their size under such conditions. Nevertheless the ability to ingest even small amounts of such material could extend the survival times of such animals. Tests conducted by Dr. J. Heinbokle and Mr. Ken King also demonstrated that large lobate Bolinopsis could ingest larvaceans and ciliates respectively, although again, nothing is known of the quantitative significance of this ability.

Food Consumption and Growth Rate and Fecundity

Reeve and Walter (1976) computed the food consumption ability of the total ctenophore population in a CEE during the 1975 experiment by estimating "water clearance" rates for various sizes of Pleurobrachia.   This was done in a manner similar to that employed for copepods feeding herbivorously, by counting the food items before and after a measured feeding period, and employing an exponential

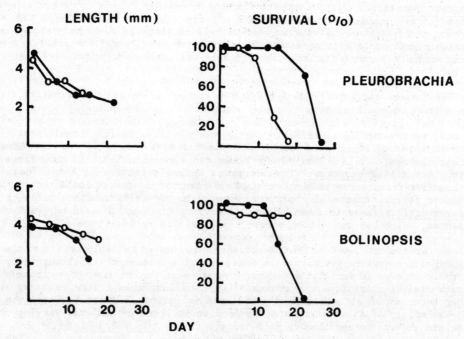

Fig. 6. Effect of starvation on length (left) and survival
         (right) of Pleurobrachia (upper) and Bolinopsis
         (lower).

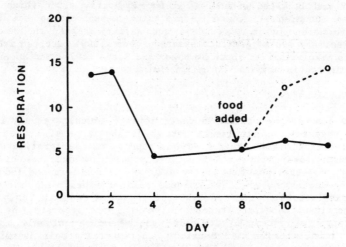

Fig. 7. Effect of starvation on respiration (in relative
         units) of Pleurobrachia. Starvation began on day
         2. Some animals (dotted line) were provided with
         food from day 8 onwards.

function to estimate the volume searched, assuming that all food items contacted were ingested. Volumes searched ranged from 5-1,150 ml day^{-1} from the smallest to largest size classes. Applying these data to population estimates, a maximum search volume at the population peak was 36 l m^{-3} day^{-1}, or 3.6 of the total CEE volume daily. Taking Bolinopsis into account as well that figure would have approximately doubled. This would have been an underestimate because larval ctenophores were not being adequately sampled at the time. A more realistic value might have been closer to 10%.

Sullivan, Reeve and Walter (unpublished) compared this method of estimation with the more direct approach of examining gut contents of preserved Pleurobrachia. In this method the actual food consumed by the ctenophores in the CEEs was determined. It showed that laboratory experiments consistently underestimated actual feeding rates both in terms of numbers of animals consumed and also in terms of the amount of carbon consumed. This underestimate could range up to a factor of four times. The reason was that Pleurobrachia demonstrated active selection of large food items. Such computations serve to demonstrate that ctenophore biomass could indeed rapidly increase at the expense of copepod biomass. In the 1978 experiment, for example, Pleurobrachia biomass increased from 0.3 to 5 mg C m^{-3} in 20 days to equal that of copepods, which had been reduced from a peak of 100 mg C m^{-3}.

Growth rates of ctenophores in CEEs cannot be estimated reliably even when the pulse of reproduction is fairly narrow in CEEs. This is because the high mortality rates of the young, great variability of growth rate between individuals, extremely rapid growth rate, and ability to decrease as well as increase in size, all obscure any interpretation of size-frequency distributions. Typical distributions (see Reeve and Walter, 1976) show rather stable age distributions of numbers, varying more in absolute numbers rather than in relative age peaks. Pleurobrachia grown from egg in the laboratory (Reeve and Walter, 1976) with unlimited food attained a length of 8 mm, at which they first start to produce eggs of their own, within 20 days at temperatures similar to those of the CEEs.

An intensive series of fecundity tests were performed during the peak of reproductive activity of 1978 in the case of Pleurobrachia. Animals were withdrawn from the CEE and isolated without food for 48 hours, after which eggs or larvae produced were counted. Fecundity and size were related, as has been shown repeatedly for ctenophores (see Reeve and Walter, 1978), the fewest eggs being produced by animals 7-8 mm and the maximum (about 1,000) recorded from an animmal of 13 mm. It is likely that as shown by Baker and Reeve (1974) for Mnemiopsis, such egg production can continue if the animal is well fed.

CONCLUSION

The CEPEX enclosures are unique in their ability to culture very large numbers of zooplankton even of the extreme size of older Bolinopsis. Ease of access, calm sampling conditions and freedom from population immigration and emigration permitted much closer observation of the population dynamics of these delicate carnivores than is ever possible in the natural environment, and led to improvements in collecting and experimental techniques. Experiments over five years confirmed laboratory predictions that very large populations of copepods inhibited rather than enhanced the likelihood of ctenophore population outbursts, by inhibiting the development of their larvae. On the other hand, relatively modest levels of herbivores, in distinction to laboratory experimental models, permitted ctenophore blooms. In some respects use of such terms becomes semantic. In absolute terms, mortality of ctenophore populations in CEEs was always at a very high level even when they bloomed, and in those terms they could be considered to be rather unsuccessful. In practice, however, as in the 1978 experiment, they remained dominant, controlling the copepod population over the entire period after the first 20 days.

ACKNOWLEDGEMENT

I sincerely appreciated the efforts of the colleagues and friends who worked with me directly on ctenophore populations over this period. They included Drs. John Gamble, "Tom" Ikeda and Barbara Sullivan, and Maryann Walter, Karen Darcy, Karen Stanlaw and Greg Louttit. For their efforts in copepod population analysis, as well as frequent assistance in this work, it is a pleasure to thank Dr. George Grice and Victoria Gibson. Thanks also to major coordinators of the CEPEX effort, Drs. David Menzel and Timothy Parsons and to all my colleagues over these years at the CEPEX site. These studies were supported by grants from the Office of the International Decade of Ocean Exploration of the National Science Foundation.

REFERENCES

Baker, L.D., and M.R. Reeve (1974). Laboratory culture of the lobate ctenophore Mnemiopsis mccradyi with notes on feeding and fecundity. Mar. Biol., 26, 57-62.

Greve, W. (1972). Ökologische Untersuchungen an Pleurobrachia pileus. II. Laboratorium suntersuchungen. Helg. wiss Meers., 23, 141-164.

Greve, W. (1977). Interspecific interaction: The analysis of complex structures in carnivorous zooplankton populations. Helg. wiss Meers., 30, 83-91.

Grice, G.D., M.R. Reeve, P. Koeller, and D.W. Menzel (1977). The use of large volume, transparent, enclosed sea surface water columns in the study of stress on plankton ecosystems. Helg. wiss Meers., 30, 118-133.

Gyllenberg, G., and W. Greve (1979). Studies on oxygen uptake in ctenophores. Ann. Zool. Fennici, 16, 44-49.

Hirota, J. (1964). Quantitative natural history of Pleurobrachia bachei in La Jolla Bight. Fish. Bull., 72, 295-335.

Ikeda, T. (1977). The effect of laboratory conditions on the extrapolation of experimental measurements to the ecology of marine zooplankton. IV. Changes in respiration and excretion rates of boreal zooplankton species maintained under fed and starved conditions. Mar. Biol., 41, 241-252.

Kremer, P. (1979). Predation by the ctenophore Mnemiopsis leidyi in Narragansett Bay, Rhode Island. Estuaries, 2, 97-105.

Menzel, D.W., and J.H. Steele (1978). The application of plastic enclosures to the study of pelagic marine biota. Rapp. P.-v. Reun Cons. int. Explor. Mer, 173, 5-6.

Parsons, T.R., K. von Bröckel, P. Koeller, M. Takahashi, M.R. Reeve, and O. Holm-Hansen (1977). The distribution of organic carbon in a marine planktonic food web following nutrient enrichment. J. exp. mar. Biol., 26, 235-247.

Reeve, M.R., and L.D. Baker (1975). Production of two planktonic carnivores (chaetognath and ctenophore) in south Florida inshore waters. Fish. Bull., 73, 238-248.

Reeve, M.R., and M.A. Walter (1976). A large-scale experiment on the growth and predation potential of ctenophore populations. In G. Mackie (Ed.), Coelenterate Ecology and Behavior. Plenum Press, New York. pp. 187-199.

Reeve, M.R., G.D. Grice, V.R. Gibson, M.A. Walter, K. Darcy, and T. Ikeda (1976). A controlled environment pollution experiment (CEPEX) and its usefulness in the study of stressed marine communities. In A.P.M. Lockwood (Ed.), Effects of Pollutants on Aquatic Organisms. Soc. Exper. Biol., Seminar Series 2. Cambridge University Press, Cambridge, Massachusetts. pp. 145-162.

Reeve, M.R., and M.A. Walter (1978). Nutritional ecology of ctenophores - a review of recent research. In F.S. Russel and M. Yonge (Eds.), Advances in Marine Biology, Vol. 15. Academic Press, London. pp. 249-287.

Reeve, M.R., M.A. Walter, and T. Ikeda (1978). Laboratory studies of ingestion and food utilization in lobate and tentaculate ctenophores. Limnol. Oceanogr., 23, 740-751.

Stanlaw, K., M.R. Reeve, and M.A. Walter (In press). The larval life history of ctenophores: A review of recent research. In P. Tardent (Ed.), Proceedings of the Fourth International Coelenterate Conference. Elsevier North Holland, Amsterdam.

Takahashi, M., W.H. Thomas, D.L.R. Seibert, J. Beers, P. Koeller, and T.R. Parsons (1975). The replication of biological events in enclosed water columns. Arch. Hydrobiol., 76, 5-23.

OXYGEN PRODUCTION AND UPTAKE BY SYMBIOTIC *AIPTASIA DIAPHANA* (RAPP), (ANTHOZOA, COELENTERATA) ADAPTED TO DIFFERENT LIGHT INTENSITIES

A. SVOBODA AND T. PORRMANN

Ruhr-Universität Bochum, Lehrstuhl für spezielle Zoologie,
D-4630 Bochum 1, F.R.G.

ABSTRACT

Laboratory measurements of O_2 uptake and production were carried out on two stocks of symbiotic Aiptasia diaphana, cultivated for 5 months at two different levels of light intensity. After 24h of recording the O_2 exchange in the tank of their upgrowth the sea anemones were transferred to the other tank and recorded there for 48h. Several symbiotic and aposymbiotic individuals were measured in the dark in a starvation and feeding program.

Both groups of sea anemones did not significantly differ in content of chlorophyll a and zooxanthellae/mg protein. The O_2 uptake rates at night were higher in the well illuminated aquarium in individuals adapted to strong light than in low light adapted ones in the low illuminated tank. Animals transferred to strong light increased their O_2 uptake and output considerably. The transfer to low light had the reverse effect. The net O_2 output per µg chlorophyll at low light intensity was higher in low light adapted animals than in strong light adapted ones at low light intensity. Low light adapted sea anemones exceeded the O_2 output of strong light adapted ones at strong light intensity. The P net/R ratio (24h mean value) exceeded 2 (balance of the energy budget) in both stocks when measured in the tank of their upgrowth, but dropped below 2 in strong light adapted individuals at the low light intensity. Starved symbiotic and aposymbiotic Aiptasia increased the O_2 uptake considerably after beeing fed in the dark.

The two different light intensities only effected the efficiency of photosynthesis by biochemical adaptation. The increased O_2 uptake rate of both stocks of sea anemones after exposure to strong light seems to result from increased feeding on photosynthetic products and is comparable to the increase of respiration rate in starved animals after beeing fed in the dark. The lower respiration rates at low light conditions can be compared with the effect of starvation in the dark.

KEYWORDS

Symbiotic coelenterates; laboratory cultivation; light adaptation;
photosynthesis efficiency; oxygen uptake rates.

INTRODUCTION

Most experiments on oxygen production and uptake in symbiotic marine
coelenterates have been carried out in individuals transferred to
the laboratory from the sea without adequate simulation of field conditions such as food supply, water flow and light conditions (diurnal changes in intensity and spectral distribution). In situ experiments avoiding these problems tend to be time consuming and depending on weather conditions, influencing photosynthesis by shading
by clouds and water turbidity. Therefore repetetive measurements
of oxygen exchange hardly can be compared. When carried out at con-
trollable light conditions in the cultivation tank productivity
measurements in the laboratory of the species can easely be repeated
and are comparable to field measurements regarding to light adaptation of the symbiotic algae.

With exception of the sea anemone Anthopleura (Shick and co-workers,
1979) the experiments mostly were carried out on individuals differing genetically which probably may react physiologically differently. Only a few symbiotic coelenterates can be cultivated in the
laboratory for years on basis of asexual multiplication. Among
Cassiopeia species the sea anemone Aiptasia diaphana (Rapp) containing the symbiotic algae Gymnodinium microadriaticum (Freuden-
thal) is the most common species cultivated in European sea aquaria.
This species occurs in shallow water depht along the Atlantic coast,
the Mediterranean and the Red Sea. Even in poor light conditions
and with scarce food supply this anemone multiplies in large numbers
by laceration of the pedal disc.

MATERIAL AND METHODS

The strain of anemones used for the experiments was grown for over
1o years in the public aquarium of the botanical garden of Essen,
F.R. Germany at light intensities ranging from O,15-1,OmW/cm^2.
Individuals were transferred to the laboratory at Bochum and cultivated in two aereated 15o l tanks for 5 months before starting the
experiments. Both tanks were connected by tubes through which the
water circulated rapidly passing through an active charcoal filter.
The temperature was kept constantly at 24°C + O,1°. The aquaria were
illuminated in a 12h light and dark cycle (diurnal) by an Osram
HQI-L high pressure mercury vapor lamp with 4ooW output. The light
intensity in the bright illuminated aquarium was about 1,9mW/cm^2 and
in the low illuminated one O,33mW/cm^2 or O,31 and O,055 Einstein/m^{2}h
(4OO-7OOnm range). This is about 4% and O,7% respectivly of the
light intensity at sea level at noon at mediterranean latitudes.
Several individuals were kept within the same circulation system in
complete darkness to bleach the zooxanthellae. All anemones were
fed twice a week with freshly hatched nauplii of Artemia salina.

Initially the anemones were starved for 24h, detached from the
aquarium glass bottom by a razor blade and placed into a small sterilized glass dish which was inserted in a glass beaker placed in
the cultivation tank. Normally, within an hour the anemones crawled
away from the sheet of mucus and algae and settled on the dish so
that the debris could be removed. After retraction of the acontia
and full expansion the anemones were placed 3 hours later in the
freshly sterilized respirometer chamber installed at the bottom of
the culture tank.

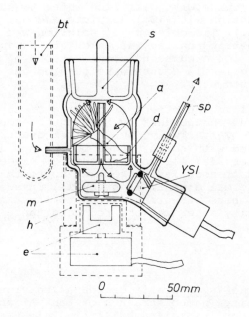

Fig. 1. Respirometer unit.
a. Aiptasia in experiment, bt. bubble trap chamber, d. dish with
Aiptasia attached, e. electric motor with driving magnet, h. perspex
cylinder for attachment of respirometer, m. stirrer magnet, s. stopper with handle, sp. tube to suction pump, YSI. oxygen electrode,
solid arrows: internal water flow, dotted arrows: flushing water
flow.

In each experiment a second respirometer chamber was run as a control. The glassblown 70ml respirometers were fitted with a ground
stopper and ground neck for the YSI oxygen electrode. The water in
front of the electrode membrane was moved by a magnetically driven
teflon bar stirrer below the inserted dish. A glass tube ran through
the center of the dish and connected the upper part of the chamber
for circulation of the content. At regular intervals the water of
the chamber was sucked out by a centrifugal pump in 1-2 minutes and
fresh tank water passed into the chamber through a capillary to
avoid bad water conditions. The range of oxygen changes was kept
within the limits of 2mg/l around the saturation values of about
8mg/l. The electronics and recorders used were the same as described
for in situ experiments (Svoboda, 1978). The oxygen content of both
respiration chambers, the temperature and difference between the

oxygen partial pressures were recorded automatically on wax paper.
The latter value corrected to the control was printed in a 5 times
magnified scale to minimize the bias.

After the experiment the anemones were homogenized for 6o seconds in
isotonic NaCl solution in a 25ml Waring blender to a total volume of
2Oml. Four ml of the homogenate were digested with 1ml 1n NaOH for
3o minutes and then run through the standard Lowry procedure (Lowry
and co-workers, 1951). The protein readings were taken at 7OOnm on
a Zeiss PQM-3 photometer using bovine serum albumin as standard.
Five ml of the homogenate were strained through a glass fibre filter
(Sartorius), rinced twice with distilled water to accelerate the
following 12h extraction with 1Oml absolute acetone in the dark
(Jeffrey, pers.comm.). The chlorophyll solution was centrifuged and
afterwards a photometric reading was taken at 663 and 63onm. Chlorophyll a and c was calculated using the equations of Jeffrey and
Humphrey (1975). Algal cell counts were carried out on the homogenate with a hematocytometer.

Experiments controlling the effect of feeding on the respiration
rate were carried out in the dark by shading the respirometers with
black polyethylene covers. Twelve hours after starting the recording
the anemones were fed with small pieces of shrimps and the recording
continued for another 24h. All measurements of oxygen production and
uptake were started at noon in the tank of upgrowth of the individual and with the normal diurnal light cycle. In a set of experiments
the respiration chambers were transferred to the tank with the alternative light intensity after 24h and recorded there for another
24-48h.

RESULTS

Chlorophyll measurements. Aiptasia diaphana individuals grown at
strong light intensity (named strain A of tank A) and such ones
grown at low light conditions (named strain B in tank B) did not
significantly differ in their chlorophyll a content and a/c ratio
per mg protein. Small individuals with a protein content of about
6mg had up to 1Oµg chlorophyll a/mg protein, large animals with
20-3Omg protein only 6-5µg chlorophyll a/mg protein (Fig. 2a). The
chlorophyll a/c ratio of $3,4 \pm 0,8$ (mean $\pm$ SD) was significantly
independent of the anemones biomass and the level of light intensity. The number of zooxanthellae/mg protein in these experiments was
independent of the light adaptation, too. Both stocks showed a slight
increase of zooxanthellae number with decreasing biomass, from
$2,5 \times 10^6$ to $3,5 \times 10^6$ zooxanthellae/mg protein from large to small individuals (Fig. 2b). The chlorophyll a content per zooxanthella in
small Aiptasia individuals was two times higher than in the large
ones.

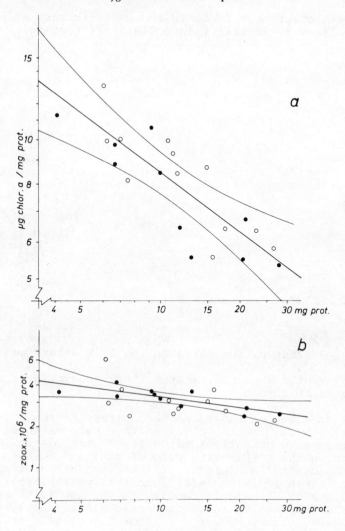

Fig. 2a. Rates of chlorophyll a/mg protein.
Fig. 2b. Rates of zooxanthellae/mg protein.
o A anemones adapted to tank A, ● B anemones adapted to tank B; the
biconcave belts indicate the 95% confidence limits of the regression.

Oxygen recordings. During the oxygen measurements the respiration
chamber used as a control continuously consumed oxygen at a constant
rate of about 1,4ug/h, independent of the light cycle. In both
Aiptasia strains there was only a slight decrease of their respiration rate (less than 1o%) when they were measured at their normal
diurnal light cycle at night. They were exposed to the light intensity at which they were cultivated and to oxygen conditions close
to saturation. Starved anemones which were fed in the dark showed a
considerable increase of the uptake rate after feeding (Fig. 3).
Maximum uptake rates in symbiotic individuals were attained 1 - 2h

after ingestion of protein food, in aposymbiotic ones after 4 - 8h.
The rates declined to initial values 24h after feeding.

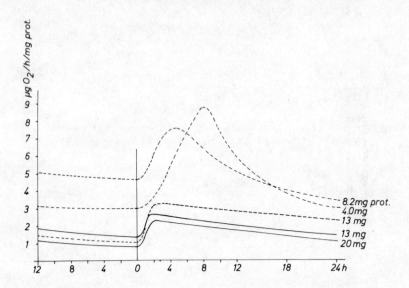

Fig. 3. Respiration rates in the dark after feeding and
            starvation.
Full lines-symbiotic anemones, dotted lines-aposymbiotic anemones;
feeding occured at time O.

Both strains of anemones showed different respiration rates at night
in the normal diurnal light cycle when they were measured in the
tank where they grew normally. Medium sized strain A individuals in
tank A had twice the respiration rate of strain B ones in tank B
(Fig. 4a). The maximal difference of the respiration rates was observed in the large individuals of both strains; the small ones,
however, did not differ significantly. In the low light adapted
strain (B) the respiration rates increased significantly with decreasing biomass, which, however, was not the case in the strong
light (A) adapted ones. When strain B animals were transferred to
tank A for 12h their respiration rate increased considerably in the
following night (25% ± 19, mean ± SD) but was still below that of
the A strain (Fig. 4b). When strain A animals in the opposite were
exposed to low light their respiration rates dropped by 21% ± 14
(mean ± SD) but larger animals at least exceeded that of the B
strain (Fig. 4c).

Net photosynthesis in both groups differed considerably depending
on the light intensity of the cultivation tank and on the light condition during the experiment. Strain A animals measured in tank A
produced about twice as much O_2/μg chlorophyll compared with individuals of strain B in tank B (Fig. 5a). After transfer of strain B
animals to tank A the photosynthesis rate increased by 175% ± 52
(mean ± SD) and exceeded even the rate of strain A anemones within
the same tank (Fig. 5b). After another diurnal cycle, however, these
high values slightly dropped (9% ± 2,8, mean ± SD) and remained constant at least for one week. When strain A animals were treated re-

versely their photosynthesis dropped by 68% ± 4 (mean ± SD) of their
strong light values and below that of B anemones in the same tank
(Fig. 5c). On the next cycle these low values slightly increased by
12% ± 3,5 (mean ± SD) but did not show further changes during one
week.

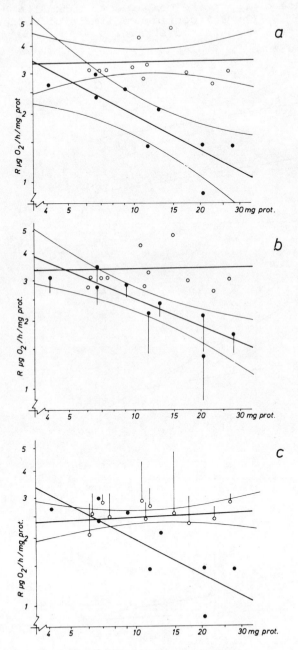

Fig. 4. Respiration rates (night).
a. o A anemones measured in tank A, ● B anemones measured in tank B;
    the biconcave belts indicate the 95% confidence limits of the re-
    gressions.

b. o A anemones measured in tank A, ● B anemones transferred and
   measured in tank A; vertical lines indicate increase of O_2 uptake
   compared to tank B; the 95% confidence limits of the regression
   are only marked for the transferred strain.

c. o A anemones transferred and measured in tank B, vertical lines
   indicate decrease of O_2 uptake compared to tank A; ● B anemones
   measured in tank B; the 95% confidence limits of regression are
   only marked for the transferred strain.

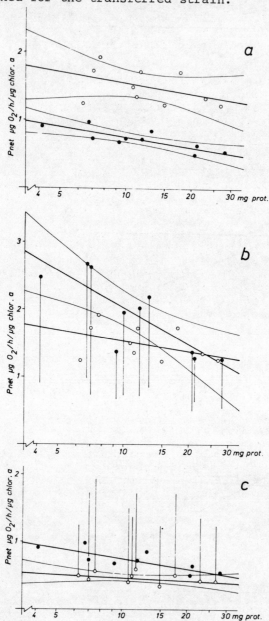

Fig. 5. Net production rates.

a. o A anemones measured in tank A, ● B anemones measured in tank B;
   the biconcave belts indicate the 95% confidence limits of the
   regressions.
b. o A anemones measured in tank A, ● B anemones transferred and
   measured in tank A; vertical lines indicate increase of O_2 production compared to tank B; the 95% confidence limits are only
   marked for the transferred strain.
c. o A anemones transferred and measured in tank B, vertical lines
   indicate decrease of O_2 production compared to tank A; ● B anemones measured in tank B; the 95% confidence limits of regression
   are only marked for the transferred strain.

Similar changes of the (net) P/R rate resulted from transfer experiments to light conditions to which they were not adapted. Strain A
anemones in tank A showed increasing P/R rates from 2-6 from large
to small individuals, strain B anemones in tank B however only slight
increasing values from 2-3 in the same range of size. The proposed
value of the equilibrium of the energy budget of 2 was only slightly
exceeded in the latter strain (Fig. 6a). When strain B anemones were
transferred and measured in tank A the P/R values increased by
167% $\pm$ 65 (mean $\pm$ SD) thus approaching the values of small A anemones in tank A and exceeding the values of the large ones by 2,5
times (Fig. 6b). After one more day in tank A the P/R rates dropped
by 8% $\pm$ 4,8 (mean $\pm$ SD) and remained constant on this level for more
than one week. When strain A animals were exposed to tank B the P/R
ratio dropped to 36% $\pm$ 6 (mean + SD) of the values they showed in
tank A and to rates below that of B animals in tank B. The large individuals consumed two times more oxygen than they produced and even
small ones remained below the maintenance level of 2. A slight increase of the P/R rate of 11% $\pm$ 6 (mean $\pm$ SD) could be recorded
after one more day of exposure in tank B which did not change for
one more week (Fig. 6c).

        DISCUSSION

The results obtained in Aiptasia diaphana regarding the constant
amount of chlorophyll and zooxanthellae under greatly varying light
conditons remain in contradiction to former Aiptasia investigations
by Steele (1976) and similar studies in corals by Drew (1972) who
claimed to have found decreasing content of zooxanthellae when the
light intensities decreased. It could be proved by Svoboda (1979)
that the mediterranean hydroids Aglaophenia tubiformis and A. harpago
considerably increase their zooxanthellae number with increasing
depth as is the case in the Red Sea coral Stylophora pistillata
(Svoboda, unpublished). On the other hand the content of zooxanthellae and chlorophyll remains constant in the Red Sea alcyonarian
Heteroxenia fuscescens (Svoboda, unpublished) independent of the
depth. Thus, Aiptasia diaphana has either the same type of adaptation or, more probably, the experiments accidentally were performed
at the lowermost possible adaptation. Further studies might reveal
such mechanisms at higher light intensities. The decrease in
zooxanthellae numbers in Steele's experiments with Aiptasia tagetes
might be due to the extreme low light intensities of 1-2% of the
sea surface conditions for individuals growing on the shallow reef
just (2 weeks) before the experiments. Furthermore, in contrast to
Steele's animals, the strains of Aiptasia diaphana originated from

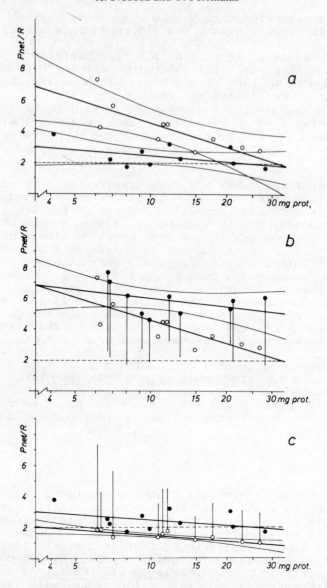

Fig. 6. Net productivity / respiration (night).
a. o A anemones measured in tank A, ● B anemones measured in tank B;
    the biconcave belts represent the 95% confidence limits of the
    regressions; the dashed line indicates the P/R ratio of 2,
    suggested as equilibrium of the energy budget.
b. o A anemones measured in tank A, ● B anemones transferred and
    measured in tank A; vertical lines indicate the increase of P/R
    ratio compared to tank B; the 95% confidence limits are only
    marked for the transferred strain.
c. o A anemones transferred and measured in tank B, ● B anemones
    measured in tank B; vertical lines indicate the decrease of P/R
    rate compared to tank A; the 95% confidence limits are only
    marked for the transferred strain.

strains which were cultivated for years under light conditions below those applied during the experiments $(0,15-1,0mW/cm^2)$.

Field measurements using large bell jars (Svoboda, 1978) showed no noticeable O_2 decrease in spite of plankton included during the experiments. However, in the present experiment under laboratory conditions using small bell jars O_2 consumption also occured in the control jar; thus in experiments with small anemones the O_2 data had to be corrected. The reason for the O_2 consumption in small bell jars is assumed to be due to bacterial respiration plus the uptake of the electrode.

During the daytime symbiotic anemones continuously receive carbohydrates and aminoacids from their algea (Muscatine and Hand, 1958; von Holt and von Holt, 1968; Taylor, 1974). When starved during oxygen recording the energy budget approximately can be calculated from the gross production of O_2 and the O_2 uptake at night. When measured in the diurnal cycle the respiration rates remain almost constant during a 12h night. The respiration and production values obtained by measurements over several days yielded always similar results. Probably a part of the fats and aminoacids stored during the day are consumed at night. When fed on protein the respiration rate of aposymbiotic individuals increases considerable by metabolizing aminoacids of the food. Concluding from the few data obtained the respiration of symbiotic anemones seems to be a little less effected. A similar increase of oxygen uptake in sedentary organisms after feeding was found in starved mussels by Bayne and co-workers (1973), although it occurs much faster in Aiptasia. The increase of the respiration rate at increased light intensity and their decrease at reduced illumination shows that the changing amount of photosynthesis has the same effect as feeding. Low yields result in low respiration rates, high yields in high ones.

The increase of respiration rates with decreasing biomass as observed in strain B is well known and may be explained by the increasing surface to volume ratio in smaller individuals. A further increase of steepness of the regression slope may be due to the fact that small individuals contain relatively more chlorophyll than larger ones and therefore get more photosynthetic products which increase the rate of metabolism. On the other hand in larger individuals there might be an influence due to self shading of zooxanthellae thus reducing photosynthesis and subsequently the metabolic rates. However, there is no explanation for the size independent respiration rate in strain A animals in tank A. It is possible that in spite of self shading of zooxanthellae there is a less reduced photosynthesis because of the higher light intensity. The higher amount of food available from this source may accelerate the metabolism in the large individuals.

Photosynthesis rates do not rise proportional to the experimentally used light intensities but they well agree with measurements of Halldal (1968) on symbiotic algae of the coral Favia. Zooxanthellae of anemones cultivated at low light intensities show an adaptation towards lower saturation light intensity and therefore succeed the production rates of the ones adapted to strong light at poor light conditions. The effect is well known from sun and shade adapted terrestrial plants and hermatypic corals (Wethey and Porter, 1976; Svoboda, 1978). Photosynthesis and respiration change rapidly with

changing light conditions but except for the small step towards
adaptation regularly observed 2 days after the transfer experiments
no further adaptation could be recorded within the following week.
Complete adaptation seems to take at least several weeks and could
not yet be measured significantly.

It is difficult to determine the maintenance level of the P/R ratio
since quantitative data of the transfer rates from zooxanthellae
to the host and metabolic rates of algae and host are not available
(McCloskey and co-workers, 1978). The transfer rates calculated
from $^{14}CO_2$ incorporation in the light of about 4o-5o% (Von Holt and
Von Holt, 1968) are only the minimum transfer rates ignoring the
^{14}C losses by algae and host respiration, the metabolic costs of
ammonium recycling and acetate recycling (Blanquet and co-workers,
1979). Extrusion of large zooxanthellae numbers was never observed
during cultivation and experiments in the laboratory; with exception of Phyllactis flosculifera they may be due to stress situations
in Steele's and Goreau's (1977) experiments. If there is no considerable waste of nutrients by leaching which seems to be in oppo-
sition to the high uptake capability of nutrients observed by
Schlichter (1975, 1978) the maintenance level may be around a P
net/R value of 2, as suggested for the experiments.

These preliminary experiments show that Aiptasia diaphana may become a useful organism in the investigation of regulation both of
contribution of zooxanthellae to nutrition of the host and adaptation
of the hosts metabolism to the rate of photosynthesis.

REFERENCES

Bayne, B.L., Thompson, R.J., and J. Widdows (1973). Some effects of
    temperature and food on the rate of oxygen consumption by Mytilus
    edulis L. In W. Wieser (Ed.), Effects of temperature on ectother-
    mic organisms. Springer, Berlin, Heidelberg and New York, pp. 181-
    193.
Blanquet, R.S., Nevenzel, J.C. and A.A. Benson (1979). Acetate incor-
    poration into lipids of the anemone Anthopleura elegantissima and
    its associated zooxanthellae. Mar.Biol., 54, 185-194.
Drew, E.A. (1972). The biology and physiology of alga-invertebrate
    symbioses. II. The density of symbiotic algal cells in a number of
    hermatypic hard corals and alcyonarians from various depths.
    J.exp.mar.Biol.Ecol., 9, 71-75.
Halldal, P. (1968). Photosynthetic capacities and photosynthetic
    action spectra of endozoic algae of the massive coral Favia.
    Biol.Bull., 134, 411-424.
Jeffrey, S.W., and G.F. Humphrey (1975). New spectrophotometric
    equations for determination chlorophylls a,b,c_1 and c_2 in higher
    plants, algae and natural phytoplankton. Biochem.Physiol. Pflanzen,
    17, 191-194.
Lowry, O.H., Rosebrough, N.J., Farr, A.L., and R.J. Randall (1951).
    Protein measurements with the Folin phenol reagent. J.biol.Chem.,
    193, 265-275.
McCloskey, L.R., Wethey, D.S., and J.W. Porter (1978). The measure-
    ment and interpretation of photosynthesis and respiration in reef
    corals. Monogr.Oceanogr.Methodol. (UNESCO), 5, 379-395.

Muscatine, L., and C. Hand (1958). Direct evidence for the transfer
   of material from symbiotic algae to the tissues of coelenterates.
   Proc.Nat.Acad.Sci.U.S.A., 44, 12,1259-1263.
Schlichter, D. (1975). The importance of dissolved organic compounds
   in sea water for the nutrition of Anemonia sulcata Pennant (Coelen-
   terata). In H. Barnes (Ed.), Proc. 9th Europ.Mar.Biol.Symp., 1975,
   Aberdeen Univ.Press, pp. 395-4o5.
Schlichter, D. (1978). On the ability of Anemonia sulcata (Coelen-
   terata, Anthozoa) to absorb charged and neutral amino acids si-
   multaneously. Mar.Biol., 45, 2,97-1o4.
Shick, J.M., Brown, W.I., Dolliver, E.G., and S.R. Kayar (1979).
   Oxygen uptake in sea anemones: effects of expansion, contraction,
   and exposure to air and the limitations of diffusion. Physiol.
   Zool. 52, 1,5o-62.
Steele, R.D. (1976). Light intensity as a factor in the regulation
   of the density of symbiotic zooxanthellae in Aiptasia tagetes
   (Coelenterata, Anthoza). J.Zool.Lond., 179, 387-4o5.
Steele, R.D., and N.I. Goreau (1977). The breakdown of symbiotic
   zooxanthellae in the sea anemone Phyllactis (= Oulactis) flosculi-
   fera (Actiniaria). J.Zool.Lond., 181, 421-437.
Svoboda, A. (1978). In situ monitoring of oxygen production and
   respiration in cnidaria with and without zooxanthellae. In D.S.
   McLusky and A.J. Berry (Eds.), Physiology and Behaviour of Marine
   Organisms. Pergamon Press, pp. 75-82.
Svoboda, A. (1979). Beitrag zur Ökologie, Biometrie und Systematik
   mediterraner Aglaophenia Arten (Hydroidea). Zool.Verh.Leiden,
   167, 1-114.
Taylor, D.L. (1974). Nutrition of algal-invertebrate symbiosis.I.
   Utilization of soluble organic nutrients by symbiont-free host.
   Proc.R.Soc.Lond.B., 186, 357-368.
Von Holt, C., and M. von Holt (1968). Transfer of photosynthetic
   products from zooxanthellae to coelenterate hosts. Comp.Biochem.
   Physiol., 24, 73-81.
Wethey, D.S., and J.W. Porter (1976), Sun and shade differences in
   productivity of reef corals. Nature, 262, 5566,281-282.

EXPANSION-CONTRACTION BEHAVIOR IN THE SEA ANEMONE *METRIDIUM SENILE:* ENVIRONMENTAL CUES AND ENERGETIC CONSEQUENCES

R. E. ROBBINS AND J. M. SHICK

Department of Zoology, University of Maine, Orono,
Maine 04469, U.S.A

ABSTRACT

Field observations on a population of *Metridium senile* indicate that a greater percentage of the anemones is expanded when the tide is running than at slack water.
Observations of anemones subjected to different current velocities and food concentrations in a laboratory flow tank reveal that current has the greatest effect on
tentacle expansion and column elongation, and that the interaction of current and
food is also highly significant. Food alone is scarcely effective, and the
response of anemones to it depends on their expansion state prior to the start of
the experiment. The role of current as the principal cause of expansion in
Metridium is discussed in terms of the animal's body plan and what is known of its
sensory receptors. Energetic considerations suggest that this current-related
behavior maximizes the amount of food-carrying water filtered per unit time and
minimizes maintenance costs, since expanding to assume the feeding posture causes
an increased metabolic rate. The energy savings may be manifested in the high
growth efficiencies in sea anemones relative to actively-swimming coelenterates.

KEYWORDS

Metridium senile; sea anemone; feeding behavior; tidal currents; functional morphology; ecological bioenergetics; growth.

INTRODUCTION

Pantin's (1965) description of animals as "predatory behavior machines" was particularly applied to cnidarians, and feeding biology has been a focal point in the
study of this phylum of lower metazoans. A notable form of behavior in the
Anthozoa is their periodic alternation between the state of expansion and feeding
and that of contraction and relative inactivity.

This expansion-contraction behavior in *Metridium senile* (L.) was poetically
described by Gosse (1860); Parker (1917, 1919), in a series of anecdotal accounts,
emphasized environmental factors influencing it. Batham and Pantin (1950a, 1950b,
1954) held that phases of expansion and contraction in *Metridium* occur spontane-
ously but can be modified by exogenous stimuli to meet specific needs. As valuable
as the older studies have been, they have rarely considered the significance of the

observed activity to the anemone in its natural habitat, and many of the stimuli employed are hardly environmentally relevant.

Metridium senile is a passive filter feeder which preys primarily on zooplankton (Purcell, 1977), relying on water currents in its tidal environment to carry food to its tentacles. Koehl (1976, 1977a, 1977b) has demonstrated the compositional, anatomical, and mechanical adaptations of *M. senile* that serve not only to enhance the food-carrying effects of the water flow but also to withstand the drag forces exerted by it. Recently, Shick, Hoffmann, and Lamb (1979) and Shick and Hoffmann (1979) have shown a relationship among flow velocities encountered in the environment, individual body size and vegetative proliferation, and possibly genotype, in a population of this species. Current, then, is central to the biology of *M. senile*, and Parker (1919) indeed showed current to be a stimulus for expansion, although his experiments were not quantitative.

It is by now well-established that expansion results in an increase in oxygen uptake in *M. senile*, due to changes in the area and thickness of the gas exchange surface (Shumway, 1978; Shick and co-workers, 1979). As *Metridium* can feed only when it is expanded, sufficient food energy must be obtained to offset the increased energy use when expanded and still provide energy for growth and reproduction. This could be accomplished by expanding primarily during periods of water movement, thus maximizing the amount of food-carrying water filtered per unit time, and minimizing metabolic rate when these conditions do not prevail. Casual field observations appeared to support this hypothesis: anemones having the body column elongated and tentacles fully expanded were seen more frequently when the tide was running than during periods of slack water.

This paper deals with the respective roles of current and food in controlling expansion state. Because *Metridium senile* does not harbor zooxanthellae, a bioenergetic and nutritional interpretation of expansion-contraction behavior is less complicated than in symbiotic species studied by Sebens and DeRiemer (1977) and Lasker (1979), in which the metabolic cost of expansion must be balanced against an increased photosynthetic input when the polyp is expanded.

MATERIALS AND METHODS

Field Studies

Preliminary observations of anemones in the field and in the laboratory were used to construct a scale of 12 behavioral states which includes both tentacle expansion and column elongation (Fig. 1), since these activities occur more or less independently. Lasker (1979) used a similar scale, but which reflected only tentacle expansion, in a study of reef corals.

A group of 33 *Metridium senile* was observed at Salt Pond, Blue Hill Falls, Maine (44°22'N, 68°34'W) for a 6-h period during an ebbing tide on 7 August 1979. Simultaneous current velocity measurements were made with hand-held propellor-type meters (Rigosha Co., Inc.). Seston collections were made using a 12-cm diameter (about the size of the expanded tentacular crown of a large anemone) plankton net held adjacent to the anemones.

Current and Food Experiments

Anemones were collected at Salt Pond from depths of 0.5 to 1.5 m at low tide. They were maintained under constant illumination without feeding or exposure to current for six weeks in a large tank of 30°/ooS seawater at 15°–16°C.

A   B   C

1

2

3

4

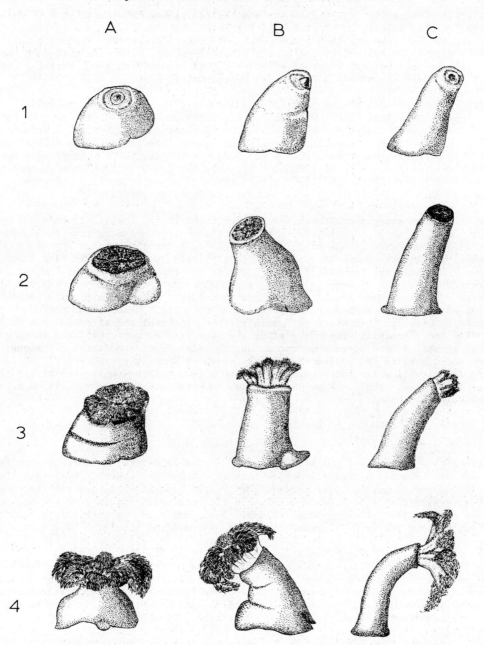

Fig. 1. Behavioral states in *Metridium senile*. Tentacle expansion: (1) tentacles fully contracted and oral disk not visible; (2) tentacles fully contracted but oral disk visible; (3) tentacles partially expanded; (4) tentacles fully expanded. Column extension: (A) column length ≤ width; (B) column length 1.0-1.5 times width; (C) column length > 1.5 times width. Drawings not to same scale.

A circular fiberglass tank 0.91 m in diameter and filled to a depth of 0.66 m
served as both holding and test facility. Water was siphoned into an external filter and chiller, and was then pumped back into the circular tank through a valve
and by-pass which controlled the amount of water entering the tank through a vertical pipe perforated with a series of holes. This produced an even, circular current, the speed of which was controlled by the valve.

Fifteen anemones, all having pedal disk diameters of at least 3 cm and well-
developed, lobed oral disks, were used. After the anemones had been in the test
tank for one week with daily feeding, they were observed for several periods of no
current and no food to establish a baseline. These observations were made after
the pump returning water from the chiller via the by-pass, which generated a non-
directional current of about 1 cm/sec, had been off for 0.5 h. Data were recorded
every 10 min for a period of 45-60 min; a total of 690 observations (46 observations on each of 15 anemones) were made over a five-day period.

For subsequent experiments three levels of current [zero, low (12-15 cm/sec), and
high (25-30 cm/sec)], corresponding to velocities experienced by this population in
the field, and three levels of food [zero, low (230 freshly-hatched *Artemia salina*
nauplii per liter), and high (460 nauplii per liter)] were tested in nine combinations. The pump was turned off for 2 h before the experiments began, and the temperature in the tank did not rise more than 0.5°C during this time. Observations
lasted 2 h (except in several cases where they lasted only 110 min) with the behavioral state of each anemone being recorded every 5 min. Observation periods were
spaced as close to 24 h apart as feasible. When the test condition was "no food"
or "low food," nauplii were added after the observations were completed to make the
total amount of food given the same as during "high food" conditions. Anemones
were not otherwise fed between tests. A length of hose by-passed the filter so
that the nauplii would not be removed during a test. Three replicates of the series of nine test conditions were performed, and the order of the treatments in each
replicate was random.

Oxygen and Glycine Uptake

Rates of oxygen uptake ($\dot{V}_{O_2}$) at 15°C and 30°/ooS were determined for expanded and
contracted anemones as described elsewhere (Shick and co-workers, 1979) with the
following modifications. The range of dry weights was slightly greater than previously, extending from 0.02 to 6.49 g. A YSI model 5720 oxygen probe, which provided more uniform stirring than the model 5420, was used, as were larger respirometry vessels (up to 2 l), which permitted near-maximal expansion by the larger
anemones.

The uptake of ^{14}C-glycine from a concentration of 0.48µM in 30°/ooS seawater at
15°C was determined in expanded and contracted small anemones. Solubilized tissues
were assayed for accumulated radioactivity on a Packard model 3255 liquid scintillation spectrometer using standard techniques. Anemone wet weights were converted
to dry weight after measuring a water content of 82% in other small specimens.

RESULTS

Field Studies

Current velocity, seston availability, and the percentage of anemones assuming
each behavioral state (Fig. 1) as related to stage of the tide are given in
Table 1. The zooplankton component of the seston consisted mainly of copepods and
nauplii.

TABLE 1 Field Observations on *Metridium senile* at Blue Hill Falls, Maine

Percentage of 33 anemones assuming various behavioral states shown in Fig. 1, with simultaneous measurements of water temperature, depth, current velocity and seston availability on an ebbing tide on 7 August 1979

Time	Temp. (°C)	Depth (m)	Current Velocity (cm/sec)*	Seston (mg dry wt./5 min)	Behavioral State											
					1A	1B	1C	2A	2B	2C	3A	3B	3C	4A	4B	4C
1150	15.0	2.0	6.7	5.0	54.5	0	0	6.1	0	0	0	0	12.1	18.2	0	9.1
1215	15.0	1.8	2.0	4.4	54.5	0	0	9.1	0	0	0	0	12.1	15.2	0	9.1
1235	15.6	1.7	2.0	-	54.5	0	0	9.1	0	0	0	0	12.1	15.2	0	9.1
1305	16.1	1.5	15.8	13.5	48.5	0	0	12.1	0	0	0	0	0	18.2	0	18.2
1330	16.1	1.5	29.2	11.7	12.1	0	0	9.1	0	0	0	18.2	0	0	12.1	48.5
1400	16.7	1.4	34.7	20.8	3.0	0	0	6.1	0	0	0	0	0	0	18.2	72.7
1420	16.7	1.4	35.2	-	3.0	0	0	6.1	0	0	0	0	0	0	18.2	72.7
1435	17.2	1.4	27.5	13.5	3.0	0	0	0	3.0	0	0	6.1	0	0	18.2	69.7
1500	17.2	1.3	21.7	-	3.0	0	0	0	6.1	0	0	0	0	0	6.1	84.8
1530	17.5	1.1	18.3	10.4	15.2	0	0	6.1	0	0	9.1	0	0	0	0	66.7
1600	17.8	1.0	10.0	5.0	9.1	0	0	18.2	0	0	6.1	0	0	30.3	36.4	0
1630	18.0	1.0	0.7	1.4	12.1	0	0	15.2	0	0	0	0	0	66.7	6.1	0
1700	18.3	0.9	0	-	9.1	0	0	15.2	6.1	0	24.2	6.1	0	21.2	18.2	0
1730	18.3	0.8	0	0.4	9.1	0	0	48.5	0	0	18.2	0	0	12.1	6.1	0
1800	18.3	0.8	0.7	-	6.1	0	0	54.5	0	0	24.2	0	0	9.1	6.1	0

*Accurate to within $\pm$ 10%

Expansion of the tentacles and elongation of the column are clearly correlated with increasing current velocity and food availability (Table 1). However, separation of the effects of current and food on the anemones' behavior is problematic, since seston availability is significantly correlated ($P < .01$) with current velocity (Fig. 2) when only the data for 7 August are considered.

As current declines from 1530h to 1700h, the relative independence of tentacle expansion and column extension is also apparent (Table 1). During the first hour, most of the anemones change from the elongated state C to the shortened state A while keeping their tentacles fully expanded. The large percentage of anemones in intermediate states of tentacle expansion (states 2 and 3) and full column contraction (state A) after a period of negligible current at the end of the observations is noteworthy and will be discussed below.

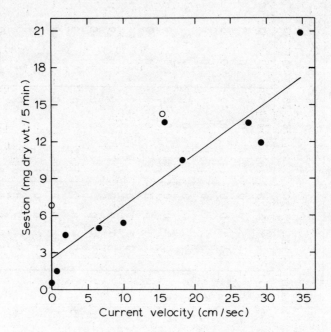

Fig. 2.   Relationship between current velocity and
          seston collected. A 12-cm diameter plank-
          ton net was held adjacent to anemones dur-
          ing ebb tide. Closed circles: collection
          of 7 August 1979. Open circles: collec-
          tion of 10 August 1979.

Current and Food Experiments

The percentage of the total "anemone time" (*i.e.*, fifteen anemones times the length of the observations in minutes) spent in each behavioral state for the baseline observations and for the total of the three replicates of each treatment is given in Table 2. There is a gradual shift from the numerically and alphabetically lower (more contracted) to the higher (more expanded) behavioral states with increasing current, and for increasing current combined with increasing food. Food alone without current has little effect; in fact, there is a shift toward the lower behavioral states under these conditions. This will be discussed below.

TABLE 2   Percent of "Anemone Time" Spent in Each Behavioral
State Under Each Experimental Condition

Current level		Zero			Low			High		
Food level		Zero	Low	High	Zero	Low	High	Zero	Low	High
	1A	48.9	70.3	69.5	22.9	38.2	28.8	10.3	8.6	5.8
	1B	0.7	0.9	0.3	0.2	0.0	0.1	0.3	0.1	0.1
	1C	0.9	0.1	0.6	0.1	0.0	0.0	0.4	0.0	0.0
	2A	8.1	0.9	7.7	13.2	8.1	12.2	15.2	9.8	7.7
	2B	2.0	1.1	0.9	1.7	0.1	0.2	0.0	0.2	0.0
Behavioral	2C	1.2	0.5	0.4	0.1	0.0	0.1	0.8	0.0	0.1
State	3A	7.4	0.9	1.3	18.9	10.3	11.1	21.0	17.8	11.9
	3B	7.5	0.8	0.2	1.9	1.2	0.8	6.7	4.9	0.7
	3C	1.9	2.1	2.0	0.6	0.1	0.3	1.5	0.1	0.7
	4A	9.6	9.9	7.4	20.6	17.4	19.5	22.8	36.8	29.9
	4B	7.8	4.2	4.3	7.6	5.5	6.1	8.2	11.8	14.2
	4C	3.9	8.7	5.4	12.2	19.1	20.7	13.2	10.0	28.8
Total number of observations		690	1050	1080	1080	1080	1080	1050	1080	1080

Table 3 presents the analyses of variance for the arc-sine transformed data. For
these analyses, as it was desired only to test for the effect of the presence or
absence of food and current, only the extremes of each variable were used (*i.e.*,
zero food, zero current; zero food, high current; high food, high current; high
food, zero current). The analyses clearly show that the presence of water flow is
the major factor affecting both tentacle expansion and column elongation. The
interaction of food and current has a significant effect on tentacle expansion.

TABLE 3   Results of Analysis of Variance for Effects of Current
and Food on Three Measures of Expansion in *M. senile*

	Tentacle State 4			Elongation State C			Behavioral State 4C		
Source	df	F	P	df	F	P	df	F	P
Replicates	2	1.374	>.25	2	4.667	>.05	2	3.597	>.05
Current	1	37.110	<.01	1	13.101	<.025	1	29.527	<.01
Food	1	3.670	>.10	1	1.178	>.25	1	4.050	>.05
Current–Food Interaction	1	10.104	<.05	1	3.023	>.10	1	4.012	>.05

The greater effectiveness of current and food together can be seen by comparing
Figs. 3 and 4. These show the number of anemones in each of the four tentacle
states immediately prior to the start of each treatment and at various times thereafter. The response in Fig. 4, representing the three replicates of high food and
high current, is quicker, larger (more anemones), and more sustained than in
Fig. 3, representing the three replicates of high current alone.

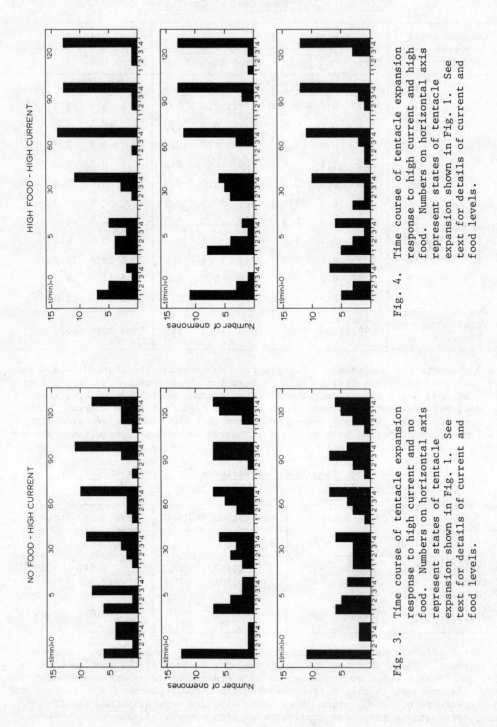

Fig. 3. Time course of tentacle expansion response to high current and no food. Numbers on horizontal axis represent states of tentacle expansion shown in Fig. 1. See text for details of current and food levels.

Fig. 4. Time course of tentacle expansion response to high current and high food. Numbers on horizontal axis represent states of tentacle expansion shown in Fig. 1. See text for details of current and food levels.

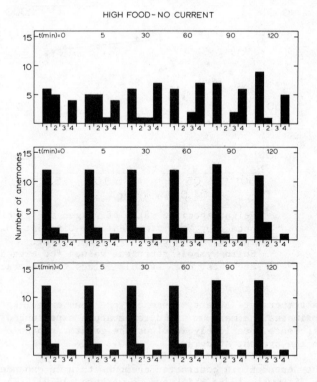

Fig. 5.   Time course of tentacle expansion
response to no current and high
food. Numbers on horizontal axis
represent states of tentacle
expansion shown in Fig. 1. See
text for details of current and
food levels.

Figure 5 presents the response of the anemones to high food alone. The response
depends on the state of the anemones prior to the addition of food -- evidently the
tentacles must be exposed for food to be detected. For example, in the second and
third replicates most of the anemones had the tentacles fully contracted (state 1),
and despite the high food concentration, showed no response to it. In the first
replicate, most anemones had tentacles at least partially expanded before the
nauplii were added, and only in this case was there any overt response to food,
although the response itself was minimal. The same effect was observed in the
three replicates of low food and no current.

Oxygen and Glycine Uptake

Weight-specific rates of oxygen uptake ($\dot{V}_{O_2}$) in expanded and contracted anemones as
functions of body weight are shown in Fig. 6. The slopes of these regressions are
virtually identical to those reported by Shumway (1978), despite the facts that our
absolute values of $\dot{V}_{O_2}$ are slightly higher and that we used a greater weight range
of anemones. [We note that the slope of -0.106 reported here for expanded anemones
is at variance with our value published earlier (Shick and co-workers, 1979). We

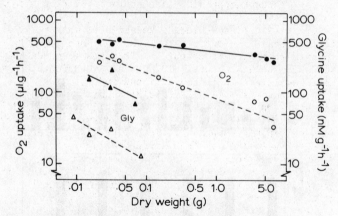

Fig. 6.   Weight-specific rates of oxygen and glycine
          uptake by *Metridium senile*. Circles: O$_2$
          uptake. Triangles: glycine uptake.
          Solid symbols represent values for expanded
          anemones, open symbols values for contracted.

cannot rule out a temperature effect, since our previous experiments were done at
22°C, which approximates the maximum field temperature experienced by this population, but the difference more likely involves the greater expansion shown by the
large anemones in the present study.]

$\dot{V}_{O2}$ is more weight-dependent in contracted anemones than in expanded ones, as discussed previously (Shumway, 1978; Shick and co-workers, 1979). Thus, contraction
will produce a proportionally greater reduction in oxygen uptake in a large anemone
than it will in a small one (Fig. 6).

Rates of glycine uptake by small specimens of *Metridium senile* are comparable to
those in other cnidarians and other marine invertebrates (reviewed by Jørgensen,
1976). Like $\dot{V}_{O2}$, glycine uptake is both diminished by contraction and is weight-
dependent, although the weight range of anemones tested is too small for a comparison of weight-dependence between expanded and contracted specimens. For small
anemones of similar weight, contraction appears to cause a proportionally greater
decrease in glycine uptake than in oxygen uptake.

       DISCUSSION

A quantitative field study confirms our subjective impression that *Metridium senile*
tends to expand when the tide is running and to contract at slack water. Because
plankton availability is directly related to current velocity, expansion during
periods of increasing water flow should increase prey capture by the anemones,
although the close association of these two environmental variables obscures the
proximate cause of the behavior.

Our laboratory experiments reveal separate and interacting roles of the environmentally relevant stimuli of water currents and planktonic prey in affecting expansion in *M. senile*. Interestingly, the presence of prey alone does not cause expansion of a fully contracted anemone. Batham and Pantin (1950b) observed that the
response of this species to homogenized mussel tissue depended on the anemone being
partially expanded when the stimulus was applied, a result we also obtained with
living prey (Fig. 5). These results likely are due to the concentration of the

putative chemo- and mechanoreceptors on the tentacles in *Metridium* (Mariscal, 1974).

The adaptations of sessile filter-feeding Cnidaria to water movement have received increasing attention recently (*e.g.*, Wainwright and Koehl, 1976). These adaptations include growth responses in colonial hydrozoans (Svoboda, 1976), gorgonians (Rees, 1972), and reef corals (Jokiel, 1978), passive orientation to current in soft corals (Leversee, 1976), and compositional and structural adaptations in a variety of forms (Wainwright and Dillon, 1969; Koehl, 1977b). Behavioral adaptations to current are less well documented (Riedl, 1971; Morin, 1976; Koehl, 1977c) and it is in this respect that *Metridium senile* is noteworthy. Even a fully contracted specimen is highly responsive to water currents, an observation dating back to the accounts of Parker (1917). Such a response is probably due to current-induced deformation of the body wall and the stimulation of the sensory cells concentrated in the parietal region of the mesenteries (Batham, Pantin, and Robson, 1960), the entire anemone acting as a giant Pacinian corpuscle (Passano and Pantin, 1955). Expansion itself is likely mediated by the SS1 conducting system known to be present in *Metridium* (McFarlane and Lawn, 1972).

Metridium senile typically occupies habitats characterized by tidal currents. In such environments, periodic increases in current velocity are good predictors of increasing food availability (Fig. 2). Thus an anemone expanding initially in response to current *per se* will expose its tentacular receptors to available prey, which must be present along with current to effect a sustained expansion response (*cf.* Figs. 3 and 4).

Given the relative paucity of differentiated sensory receptors in anthozoans (Pantin, 1965; Mariscal, 1974), it follows that those that they do have will be attuned to the critical features of the environment, the "key stimuli" of Pantin (1965). In the case of *M. senile*, which is highly adapted as a passive filter feeder (Koehl, 1976), it follows that water flow is such a stimulus. Likewise, it is understandable that the presence of planktonic prey alone is less effective, since the anemone captures it by direct interception and by inertial impaction (Rubenstein and Koehl, 1977), both of which modes require movement of the prey-carrying water across the tentacles.

Even in these lowest metazoans to possess a nervous system, the behavioral response to current and food is not all-or-none, being modifiable according to both current and prey levels (Table 2). In terms of the sensors involved, this graded response seems to be a function of increasing frequency of prey impacts, and is most apparent as the percentage of anemones assuming state 4C: in both low and high current, the response increases markedly with prey density, but not in the absence of current.

Light has been invoked as a factor affecting expansion-contraction behavior in *M. senile* (Parker, 1919; Batham and Pantin, 1950b). However, these workers' laboratory observations typically involved abrupt changes in illumination that would not occur in nature, and Parker's (1919) observations of anemones in a tide pool neglected other environmental factors. We have seen anemones in all stages of expansion during both day and night in the field, and in the laboratory have observed them to contract not only in response to sudden illumination but also to sudden darkening. *Metridium* lacks the zooxanthellae found in many other anthozoans in which light-correlated expansion has been linked to maximizing photosynthesis by the symbionts (Pearse, 1974; Sebens and DeRiemer, 1977; Lasker, 1979). Therefore we do not feel that in the field light is an important environmental cue for expansion or contraction in *Metridium* in a tidal environment, but under constant laboratory conditions this anemone will respond to it in the absence of relevant stimuli.

The question arises, why do the anemones not always expand maximally to capture most efficiently whatever prey may be available? One answer may lie in the simplicity of the animal's body plan and some related cost accounting. Extension of the column and expansion of the tentacles result in an increased oxygen consumption in *M. senile*, largely due to increases in the surface area-to-body mass ratio and decreases in diffusion distances (Shumway, 1978; Shick and co-workers, 1979). Thus, tentacles serve not only as prey capture surfaces but also as gas exchange sites, so than an increase in $\dot{V}_{O_2}$ (metabolic cost) is a passive consequence of expanding to feed. By remaining contracted when food is unavailable, *Metridium* minimizes its maintenance costs and saves its energy reserves.

The direct relationship between degree of expansion and current/prey levels (Tables 1 and 2) may also be interpretable in both anatomical and energetic terms. Batham and Pantin (1950b) have noted that different phases of activity in *Metridium* which involve the same groups of effectors may conflict with each other. A maximal expansion in response to only minimal food availability can result in a net energy loss to the animal, so that intermediate expansion states may represent compromises between the need to capture prey and that to conserve existing reserves. Slight changes in expansion state in fact result in substantial changes in $\dot{V}_{O_2}$: a 2.72 g animal changing from state 3B to state 4B in a respirometry vessel showed a 40% increase in oxygen uptake. Part of this increase may be due to initiation of irrigation of the coelenteron, as shown by Jones, Pickthall, and Nesbitt (1977).

Body size also influences expansion-contraction behavior. Small anemones (those less than 1 cm in diameter and without well-developed, lobed oral disks and frilly tentacles) spend more time expanded (68%) than do large anemones (49%) under identical conditions in laboratory tanks. Ivleva (1964) also noted that juveniles of *Actinia equina* spend more time expanded than do adults. In *Metridium*, we suggest that the small anemones must spend more time feeding due to their presumably less efficient filtering apparatus, and to their lack of access to the current mainstream (a result both of their size and of "shading" by larger anemones).

Dissolved free amino acids (DFAA) are an additional source of nutrition in sea anemones (Schlichter, 1978), and may be particularly important in less well-nourished small specimens (*cf.*, Shick, 1975), which also have a large surface-to-mass ratio. When a small specimen of *Metridium* contracts, glycine uptake is reduced by about 80% (Fig. 6), which is close to the value predicted by the decrease in surface area (Shick and co-workers, 1979). Thus, although glycine uptake sites do not appear to be particularly concentrated on the tentacular, relative to the columnar, ectoderm, a large decrease in DFAA uptake will nevertheless result from contraction in a small specimen because the greater part of its exchange surface is provided by the tentacles. Therefore, it pays a small anemone to spend more time expanded to use this source of food.

The size-related difference in expansion behavior is probably related to relative metabolic costs as well as to feeding and DFAA uptake, since contraction results in a proportionally greater energy savings in a large anemone than in a small one (Fig. 6). Although $\dot{V}_{O_2}$ is reduced by 85% when a 5-gram specimen contracts, the reduction is only 50% in a 50-milligram anemone. This decrease in the small animal is actually less than predicted from the percent of external surface area lost in contraction, so that there must be delivery of O_2 to internal tissues in a small contracted anemone, the column of which presents little resistance to diffusion. The decrease is more than that predicted from surface area reduction in a large animal, which implies that irrigation of the gastrovascular cavity (which must cease when the animal contracts) is important in the large animals with their more massive parietal musculature and mesenteries.

Although Ivleva (1964) has suggested that the small component of the animal's energy budget due to "metabolism" in *Actinia equina* is due to the near absence of an "active" metabolic rate, the fact remains that the weight-specific rates of oxygen uptake by expanded anemones are comparable to the routine rates of oxygen

uptake in other marine invertebrates of similar size when expressed in terms of dry
weight (Brafield and Chapman, 1965; Shick and co-workers, 1979). However, the substantial energy savings afforded by periodic contraction do appear to affect other
portions of an anemone's energy budget, since relative to actively-swimming medusae
and ctenophores, anemones tend to have high gross growth efficiencies (K_1, or
units of growth/units of ingested food). Ivleva (1964) reports K_1 values of 0.38
and 0.46 (depending on daily ration) for *Actinia equina*, and Wood and Shick (unpublished data) have determined an efficiency of 0.38 (range 0.29-0.54) in *Anthopleura
elegantissima*. In contrast, the hydrozoan medusa *Aequorea victorea* has a K_1 of
0.19 (Arai, 1979), and the ctenophore *Mnemiopsis mccradyi* has an efficiency of
about 0.10 (Reeve, Walter, and Ikeda, 1978). The latter authors have noted that
the higher value of 0.60 reported for another ctenophore (*Pleurobrachia*) was inap-
propriate, since it was calculated on the basis of ash-free dry weight of food and
ctenophore, and the *organic* dry weight of the latter is a much smaller proportion
of the total dry weight than occurs in its copepod prey. Similarly, the growth
efficiencies of 0.37 and 0.38 for the scyphomedusae *Cyanea capillata* and *Aurelia
aurita* (Fraser, 1969) are superficially similar to the efficiencies in sea
anemones, but they too were calculated on a dry weight basis, and these medusae
have a substantially higher inorganic ash component (about 60-70% of total dry
weight) than do anemones (about 10-20%) (Vinogradov, 1953; Shick and co-workers,
1979).

Our present data also indicate that the nutritional state of the animal affects the
expansion response. The large anemones were initially starved in the absence of
current for six weeks, then fed once daily in current. After one week, the baseline observations (no food, no current) were made. The data for the treatments of
low food-no current and high food-no current (Table 2) are substantially different,
with an increase in the percent of time spent contracted. A possible explanation
is that during the baseline observations the anemones had not fully replenished the
reserves used during six weeks of enforced starvation and were expanding a greater
portion of the time to capture whatever food may have been present. In the subsequent activities, after the anemones had been exposed daily to food and current,
which would have further replenished their energy reserves and possibly have "con-
ditioned" them to current, they remained contracted a greater percentage of the
time until cued by current to expand. When the treatment involved no current,
those which were not at least partially expanded could not detect the presence of
food (as discussed above) and therefore remained contracted.

The hypothesis that the anemones were modifying a behavior pattern established during the baseline observations is supported by the data in Table 2, in that the
increase in "anemone time" spent in tentacle state 1 when food was present occurred
at the expense of state 3 and to a lesser extent state 2, relative to the baseline
observations. Also, the increase in column elongation state A (with food present)
occurred at the expense of state B. This indicates that during the baseline study
the anemones were adopting intermediate states of expansion, compromising between
increased energy utilization and possible food capture when expanded.

Since *Metridium* can respond rapidly to conditions under which prey capture is
likely, what is the adaptive value of the spontaneous component of expansion behavior emphasized by Batham and Pantin (1950a)? An answer to this question must be
speculative. Although zooplankton is the principal food of *Metridium*, the anemone
is an opportunistic feeder and will also eat large pieces of living prey and car-
rion (Purcell, 1977). Because of the tentacular location of its chemoreceptors, a
contracted specimen of *Metridium* would not detect such prey, and periodic expansion
would allow the animal to "test" the environment for this sporadic food source.
The swaying behavior described by Parker (1919) has been suggested as an adaptation
for acquiring such food items. In this context it is informative that in *Tealia
felina*, which is predominantly macrophagous and not as dependent on water currents
for feeding, chemoreceptors are found on the column ectoderm (Lawn, 1975).

The possibility also exists that the behavior is not truly "spontaneous." Mori (1960) has shown that the ammonia concentration in the body fluid of the sea pen *Cavernularia obesa* varies inversely with degree of expansion. Periodic expansion could be a response to an internal accumulation of metabolic wastes, which would more readily be released by an expanded animal.

Many workers, notably Batham and Pantin (1950b), have remarked on the variety of behavior patterns of which sea anemones are capable despite their simplicity of organization. It is because of this simplicity that structures such as tentacles are necessarily multifunctional, and behavioral responses to a particular stimulus will have immediate physiological consequences not directly related to that stimulus. The variability in behavior patterns in *Metridium senile* likely reflects continuous compromises necessitated by conflicting physiological and bioenergetic demands and the limitations of the animal's body plan.

ACKNOWLEDGMENT

These studies were greatly facilitated by the cooperation of many individuals whose assistance it is a pleasure to acknowledge: Dr. B.D. Sidell, for advice on the flow tank design; Drs. J.D. McCleave and R.L. Vadas, for the loan of equipment; Mr. R.G. Gustafson, for assistance in field studies; Dr. J.M. Ringo, for advice on statistical analysis; and Dr. R.K. Josephson, for his critical reading of the manuscript. Travel funds were provided by the University of Maine. The research was supported by National Science Foundation grant PCM-7911027 (Regulatory Biology) to J.M.S.

REFERENCES

Arai, M.N. (1979). Growth rates of *Aequorea* medusae. In P. Tardent (Ed.), *Coelenterate Development and Morphogenesis*. Elsevier/North Holland, Amsterdam. In press.

Batham, E.J., and C.F.A. Pantin (1950a). Inherent activity in the sea anemone *Metridium senile* (L.). *J. Exp. Biol.*, 27, 290-301.

Batham, E.J., and C.F.A. Pantin (1950b). Phases of activity in the sea-anemone *Metridium senile* (L.), and their relation to external stimuli. *J. Exp. Biol.*, 27, 377-399.

Batham, E.J., and C.F.A. Pantin (1954). Slow contraction and its relation to spontaneous activity in the sea-anemone *Metridium senile* (L.). *J. Exp. Biol.*, 31, 84-103.

Batham, E.J., C.F.A. Pantin, and E.A. Robson (1960). The nerve-net of the sea anemone, *Metridium senile* (L.): the mesenteries and column. *Quart. J. Microscop. Sci.*, 101, 487-510.

Brafield, A.E., and G. Chapman (1965). The oxygen consumption of *Pennatula rubra* Ellis and some other anthozoans. *Z. vergl. Physiol.*, 50, 363-370.

Fraser, J.H. (1969). Experimental feeding of some medusae and Chaetognatha. *J. Fish. Res. Board Can.*, 26, 1743-1762.

Gosse, P.H. (1860). *Actinologia Britannica. A History of the British Sea-Anemones and Corals*. Van Voorst, London.

Ivleva, I.V. (1964). Elements of energetic balance in sea anemones. Acad. Sci. U.S.S.R. *Trans. Sevastopol Biol. Sta.*, 25, 410-428 (in Russian).

Jokiel, P.L. (1978). Effects of water motion on reef corals. *J. Exp. Mar. Biol. Ecol.*, 35, 87-97.

Jones, W.C., V.J. Pickthall, and S.P. Nesbitt (1977). A respiratory rhythm in sea anemones. *J. Exp. Biol.*, 68, 187-198.

Jørgensen, C.B. (1976). August Pütter, August Krogh, and modern concepts on the role of dissolved organic matter as food for aquatic animals. *Biol. Rev.*, 51, 291-328.

Koehl, M.A.R. (1976). Mechanical design in sea anemones. In G.O. Mackie (Ed.), *Coelenterate Ecology and Behavior*. Plenum Press, New York. pp. 23-31.

Koehl, M.A.R. (1977a). Mechanical diversity of connective tissue of the body wall of sea anemones. *J. Exp. Biol.*, 69, 107-125.

Koehl, M.A.R. (1977b). Mechanical organization of cantilever-like sessile organisms: sea anemones. *J. Exp. Biol.*, 69, 127-142.

Koehl, M.A.R. (1977c). Effects of sea anemones on the flow forces they encounter. *J. Exp. Biol.*, 69, 87-105.

Lasker, H.R. (1979). Light dependent activity patterns among reef corals: *Montastrea cavernosa*. *Biol. Bull.*, 156, 196-211.

Lawn, I.D. (1975). An electrophysiological analysis of chemoreception in the sea anemone *Tealia felina*. *J. Exp. Biol.*, 63, 525-536.

Leversee, G.J. (1976). Flow and feeding in fan-shaped colonies of the gorgonian coral, *Leptogorgia*. *Biol. Bull.*, 151, 344-356.

Mariscal, R.N. (1974). Scanning electron microscopy of the sensory surface of the tentacles of sea anemones and corals. *Z. Zellforsch.*, 147, 149-156.

McFarlane, I.D., and I.D. Lawn (1972). Expansion and contraction of the oral disc in the sea anemone *Tealia felina*. *J. Exp. Biol.*, 57, 633-649.

Mori, S. (1960). Influence of environmental and physiological factors on the daily rhythmic activity of a sea-pen. *Cold Spring Harbor Symp. Quant. Biol.*, 25, 333-344.

Morin, J.G. (1976). Probable functions of bioluminescence in the Pennatulacea. In G.O. Mackie (Ed.), *Coelenterate Ecology and Behavior*. Plenum Press, New York. pp. 629-638.

Pantin, C.F.A. (1965). Capabilities of the coelenterate behavior machine. *Am. Zool.*, 5, 581-589.

Parker, G.H. (1917). Actinian behavior. *J. Exp. Zool.*, 22, 193-229.

Parker, G.H. (1919). *The Elementary Nervous System*. Lippincott, Philadelphia.

Passano, L.M., and C.F.A. Pantin (1955). Mechanical stimulation in the sea-anemone *Calliactis parasitica*. *Proc. R. Soc. B*, 143, 226-238.

Pearse, V.B. (1974). Modification of sea anemone behavior by symbiotic zooxanthellae: expansion and contraction. *Biol. Bull.*, 147, 641-651.

Purcell, J.E. (1977). The diet of large and small individuals of the sea anemone *Metridium senile*. *Bull. S. Cal. Acad. Sci.*, 76, 168-172.

Rees, J.T. (1972). The effect of current on growth form in an octocoral. *J. Exp. Mar. Biol. Ecol.*, 10, 115-123.

Reeve, M.R., M.A. Walter, and T. Ikeda (1978). Laboratory studies of ingestion and food utilization in lobate and tentaculate ctenophores. *Limnol. Oceanogr.*, 23, 740-751.

Riedl, R. (1971). Water movement. In O. Kinne (ed.), *Marine Ecology*, Vol. I, Part 2. Wiley-Interscience, New York. Chapter 5, pp. 1123-1156.

Rubenstein, D.I., and M.A.R. Koehl (1977). The mechanisms of filter feeding: some theoretical considerations. *Am. Nat.*, 111, 981-994.

Schlichter, D. (1978). On the ability of *Anemonia sulcata* (Coelenterata: Anthozoa) to absorb charged and neutral amino acids simultaneously. *Mar. Biol.*, 45, 97-104.

Sebens, K.P., and K. DeRiemer (1977). Diel cycles of expansion and contraction in coral reef anthozoans. *Mar. Biol.*, 43, 247-256.

Shick, J.M. (1975). Uptake and utilization of dissolved glycine by *Aurelia aurita* scyphistomae: temperature effects on the uptake process; nutritional role of dissolved amino acids. *Biol. Bull.*, 148, 117-140.

Shick, J.M., W.I. Brown, E.G. Dolliver, and S.R. Kayar (1979). Oxygen uptake in sea anemones: effects of expansion, contraction, and exposure to air and the limitations of diffusion. *Physiol. Zool.*, 52, 50-62.

Shick, J.M., and R.J. Hoffmann (1979). Effects of the trophic and physical environments on asexual reproduction and body size in the sea anemone *Metridium senile*. In P. Tardent (ed.), *Coelenterate Development and Morphogenesis*. Elsevier/North Holland, Amsterdam. In press.

Shick, J.M., R.J. Hoffmann, and A.N. Lamb (1979). Asexual reproduction, population structure, and genotype-environment interactions in sea anemones. *Am. Zool.*, 19, in press.

Shumway, S.E. (1978). Activity and respiration in the anemone, *Metridium senile* (L.) exposed to salinity fluctuations. *J. Exp. Mar. Biol. Ecol.*, 33, 85–92.

Svoboda, A. (1976). The orientation of *Aglaophenia* fans to current in laboratory conditions (Hydrozoa, Coelenterata). In G.O. Mackie (Ed.), *Coelenterate Ecology and Behavior*. Plenum Press, New York. pp. 41–48.

Vinogradov, A.P. (1953). *The Elementary Chemical Composition of Marine Organisms*. Memoir 2. Sears Foundation for Marine Research, Yale University, New Haven.

Wainwright, S.A., and J.R. Dillon (1969). On the orientation of sea fans (genus *Gorgonia*). *Biol. Bull.*, 136, 130–139.

Wainwright, S.A., and M.A.R. Koehl (1976). The nature of flow and the reaction of benthic Cnidaria to it. In G.O. Mackie (Ed.), *Coelenterate Ecology and Behavior*. Plenum Press, New York. pp. 5–21.

INTRACELLULAR AND TRANSEPITHELIAL POTENTIAL CHANGES IN TENTACLE TISSUE OF SEA ANEMONES DURING THE ABSORPTION OF DISSOLVED ORGANIC MATERIAL

D. SCHLICHTER AND W. ADAM

Zoologisches Institut der Universität Köln, Weyertal 119
D 5000 Köln 41, F.R.G.

ABSTRACT

Preliminary studies were made on electric phenomena occurring in sea anemones in conjunction with the uptake of dissolved organic material. Four types of potentials could be measured: 1) transepithelial rest potentials of + 180 µV to + 350 µV, 2) transepithelial transport potentials, which were + 160 µV higher, 3) membrane rest potentials of up to − 14 mV, in average − 3,9 mV ± 1,4 mV and 4) membrane transport potentials which caused an increase of +1 to +4 mV of the rest potential. The effect of various sugars and amino acids on the rest potentials was investigated. The organic compounds resulted in the potential changes as mentioned above. The changes of the transepithelial rest potentials after adding either sugars or amino acids occurred slowly; the membrane rest potentials changed at once. A number of metabolic inhibitors were tested as to their effect on the transepithelial transport potentials. The results are compared with those obtained from various vertebrate and invertebrate tissues.

KEYWORDS

DOM absorption; sea anemone; intracellular potentials; transepithelial potentials; inhibitors.

INTRODUCTION

To date the standard methods employed to monitor the uptake of dissolved organic material (DOM) by marine and brackish invertebrates have been: 1. Measurement of the unidirectional influxes by means of radioactively labelled organic compounds, either indirectly by measuring the decreasing concentration of the labelled compound in the incubation medium or directly by determination of the incorporated labelled compound in the tissue itself. These biochemical studies do not only provide information about the substances actually taken up, but what is even more important, they also give clues about the pathways along which the absorbed compounds are channeled in. The main disadvantage of this method lies in the fact that possible effluxes

of other organic compounds remain virtually untracable. On the other
hand the tissue which actually absorbs DOM can be identified in autoradiographs and it is possible to follow the course of events taking
place during the distribution of the labelled compound within the
entire body. 2. Highly sensitive fluorometric methods have been used
to detect the net fluxes of certain organic substances. Especially
for calculations on energy flow it is advantageous to use this method
due to the fact that the variation of the total amount of e.g. amino
acids in the medium (influx and mutual efflux) can be registered.
Undenfriend and co-workers (1972) developed a method with which it is
possible to measure amino acids, peptides, and primary amines in
the picomole range. Stephens (1975) used this fluorescamine method
to determine the net fluxes of amino acids through the body wall of
annelids. Gomme (1979) also used fluorometric methods to study the
absorption of glucose in polychaets.
For further studies of DOM uptake in sea anemones an additional
method, an electrophysiological one, may now be used with some success.
As the neccessary prerequisits do now exist, it seems possible to
carry out these investigations successfully. Due to autoradiographs
of tentacle tissue from sea anemones it has been well established
that the absorption process takes place on the apical membrane of
the epidermis (Schlichter 1973, 1974). Electronmicrographs clearly
show that the tentacle epidermis of sea anemones bears a large amount
of microvilli (Schlichter 1973). Moreover histochemical investigations confirm the impression gained by autoradiographic and elec-
tronmicroscopical studies. The apical ectoderm membrane develops a
high content of alkaline phosphatases, these being enzymes typical
of membranes in which absorptive and transport processes take place
(Fig. 1).

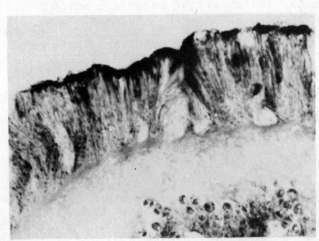

Epidermis

Mesogloea

Gastrodermis

Fig. 1. Alkaline phosphatases in high concentra-
        tion in the apical epidermis membrane of
        Anemonia sulcata; (following the method
        of Gomori-Takamatsu).

The use of electrophysiological methods for investigations concerning DOM uptake in sea anemones was very much encouraged by electrophysiological experiments of other authors studying the flux of amino
acids or glucose in the tissue of the small intestine of rats (Okada
and co-workers 1977), in kidney tubules of rats (Samarzija 1978)or in
the midgut of insects (Giordana and Sacchi 1978). Furthermore an in
depth review on the solute transport theory is available (Frömter
1979).
Using sea anemone tentacle tissue for electrophysiological studies
has the advantage over the other tissues studied that it may simply
be isolated from the sea anemones and that the isolated tentacles
as well as pieces of tentacle tissue are still in a position to absorb DOM for a considerable period of time after amputation.

MATERIALS AND METHODS

In the experiments isolated tentacles of Radianthus sp, a reef anemone or of Anemonia sulcata were used. The tentacles were amputated
from the anemones by laying a ligature around the basal part of the
tentacle behind which it was cut off. Then a second ligature was
made near the tip of the isolated tentacle in order to prevent the
opening of the tip, causing a leak in the small "tentacle sausage".
After this the tentacle was transferred to the incubation chamber
and fixed with glass needles to the bottom which was covered with
paraffin. The incubation chamber was filled with 50 ml of artificial
sea water (sterilized) at 20° C and a salinity of 35 ‰. If necessary the tentacle was incubated in a small incubation chamber (0.5
ml) connected to a reservoir of sea water and a peristaltic pump
which was used to change the incubation medium from sea water to the
test medium and vice versa. The glas microelectrodes used were pro-

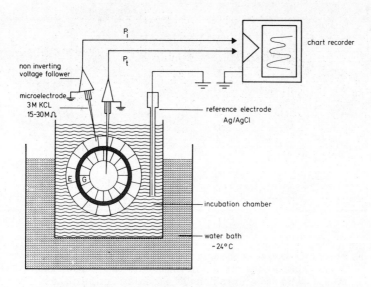

Fig. 2. The experimental set up for the measure-
        ment of transepithelial or cell potentials;
        E = epidermis; G = gastrodermis.

duced on a vertical puller adjusted for electrodes with a resistance
of 20-30 M Ohm. The microelectrodes were filled by capillary action
with 3 M KCl. After filling, the electrodes were checked for correct
resistance and minimum drift. An Ingold 373M3 reference electrode
was used. The experimental set up for the measurements is shown in
Fig. 2. The microelectrodes were coupled to a high impedance voltage
follower via a suitable microelectrode holder also filled with 3 M KCl.
A potentiometric recorder (BBC Servogor 310, with integrated ampli-
fier) was connected to the voltage follower in order to record the
potentials picked up by the electrodes. Aided by a micromanipulator
the electrodes either penetrated the epidermis, mesogloea and gastrodermis ending up in the coelenteron of the tentacle, or the microelectrodes were made only to penetrate the apical epidermis membrane;
so two types of potential were measured: Firstly a transepithelial
potential (P_t) and secondly an intracellular, membrane or cell potential (P_i).

RESULTS

Transepithelial Rest Potentials

The potential difference between the surrounding water and the coelenteron amounts to +180-350 µV inside positive, this being the trans-

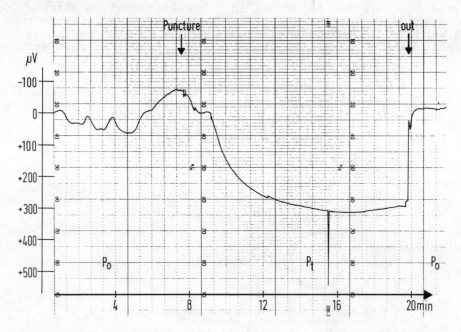

Fig. 3. Due to a temporary short circuit between
the coelenteron and the surrounding water
the transepithelial rest potential is built
up some minutes after penetrating the
tentacle wall of Radianthus sp.

epithelial rest potential. This rest potential may either be established immediately after penetration of the microelectrode through the three layers of the tentacle wall (Figs. 4, 6) or if a leak in the tissue remained, thus temporarily short circuiting coelenteron and surrounding water, it took some minutes to build up the potential to it's normal value, due to the self sealing action of the tissue (Fig. 3). Having once established a transepithelial potential it remained constant over considerable periods of time (up to 90 minutes), even in spite of movements of the tentacle or shape changes taking place.

Transepithelial Transport Potentials

A transepithelial transport potential could be evoked by changing the incubation medium from sea water to one of the following solutions containig either D-glucose, succrose, mannitol; L-isomers of aspartic acid, glutamic acid, methionine, threonine, serine, valine, alanine, leucine, isoleucine, phenylalanine, proline or glycine and D-leucine. The potentials evoked by these solutions of concentrations ranging from 30 - 100 mM, which are unnaturally high, amounted to +120 to +160 µV, independent of the compound added. Table 1 shows the changing transepithelial transport potential for different concentrations of glucose. The readings were obtained from one tentacle of Radianthus sp, starting with the low concentrations.

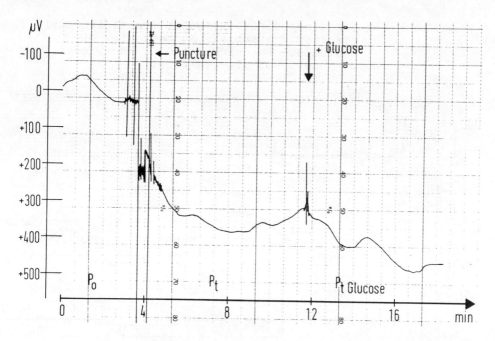

Fig. 4. Transepithelial transport potential in a tentacle of Radianthus sp, evoked by 40 mM glucose.

TABLE 1 Effect of Glucose offered in different Concentrations on the Transepithelial Transport Potential

Glucose concentration mM	Transepithelial transport-potential μV
5	0
10	0
30	+20
70	+50
100	+100

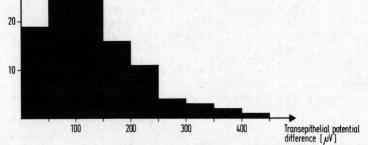

Fig. 5. Frequency distribution of transepithelial transport potentials in tentacles of Radianthus sp, evoked by 100 mM glucose. n = number of observations.

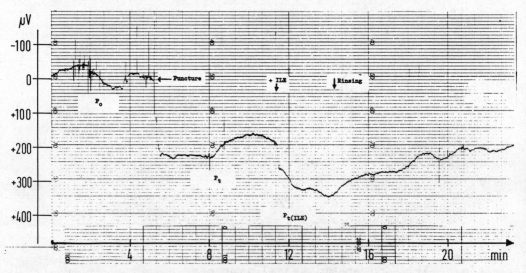

Fig. 6. Transepithelial transport potential in a tentacle of Radianthus sp evoked by isoleucine (40 mM final concentration). After rinsing with sea water the rest potential is reestablished.

Repeated measurements on a number of tentacles showed that the transepithelial transport potential evoked by 100 mM glucose amounted to +120 µV + 80 µV (n=116). The data are summarized in Fig. 5. The elevated potentials only remained as long as the tentacle was immersed in the test solution and the rest potential was reestablished when the incubation medium was changed back to sea water (Fig. 6). One single tentacle will respond to quite a few (10-15) changes of the incubation medium without any apparent reduction in response.

Influence of inhibitors on transepithelial transport potentials. A number of metabolic inhibitors were tested as to their effect on a glucose induced transepithelial transport potential. For this, the transepithelial transport potential of tentacles of Radianthus sp was first measured by exchanging sea water to 100 mM glucose, then it was changed to a solution containing glucose and one certain inhibitor for various periods of time, this was followed by a change back to glucose. Some typical results are given in Table 2.

TABLE 2 The Influence of various Inhibitors on Transepithelial Transport Potentials evoked by 100 mM Glucose.

Inhibitor-concentration	Duration of incubation min	Transepithelial transport potential before and after incubation in µV		
		before	after	effect
Phlorizine 10^{-3}M	60	100	120	0
Ouabain 10^{-3}M	30	80	100	0
"	180	110	120	0
DNP approx. 10^{-2}M	15	120	20	+
"	20	100	0	++
"	30	130	0	++
KCN 10^{-2}M	75	120	150	0

Membrane Potentials

Membrane Rest Potentials. Measurements of membrane potentials was far more difficult compared to transepithelial potentials. The main difficulties were, to fix the tentacle tissue in an adequate position and to register stable potentials over a period of more than one minute. The mean value of 135 measurements made is -3.9 mV + 1.4 mV, cell interior negative. The frequency distribution is shown in Fig. 7. Only those measurements are contained in the figure where the period of stability exceeded 90 seconds. An intracellular recording is shown in Fig. 8. In some cases it was possible to obtain a stable reading for as long as 30 minutes, but usually the potentials remained stable for 2 to 4 minutes. For short periods of time the membrane potentials very often decreased to -20 or -30 mV, but so far it has not been possible to get stable readings in this range.

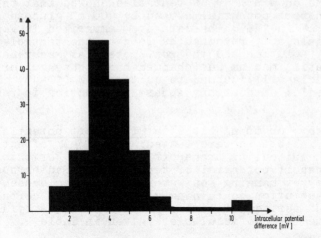

Fig. 7. Frequency distribution of membrane rest
potentials of epidermal cells of Radianthus sp. n = number of observations.

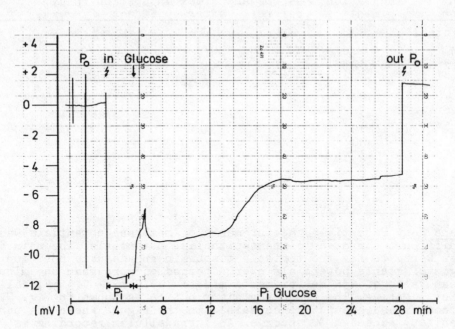

Fig. 8. During the absorption of glucose the
membrane potential is changed in epidermal cells of Anemonia sulcata; final
glucose concentration 100 µM.

<u>Membrane Transport Potentials.</u> By virtue of the experimental diffi-
culties mentioned above, till now only few transport potentials could

be measured. Fig. 8 shows the result of a successful registration.
Adding glucose to a final concentration of 100 µM, the rest potential
was elevated by + 6 mV and remained stable for a longer period. With-
drawal of the microelectrode resulted in a registration of the same
potential as before the perforation.

DISCUSSION

Transepithelial Potentials

The transepithelial potential difference appears due to the fact
that there are different numbers of differently charged particles
in the coelenteron and the sea water. When adding glucose or amino
acids to the sea water a change in the potential level results. That
the potential rise is related to the absorption of DOM is probably
due to the fact that positively charged particles are extruded from
the gastrodermis and released into the coelenteron in conjunction
with the epidermal uptake of DOM. An increase of Na^+ -ions in the
fluid of the coelenteron was not detectable by means of measurements with an atomic absorption spectrophotometer. One must state
that the change of the transepithelial potential need not neccessa-
rily be related to the uptake of DOM but may also be explained by
a Donnan equilibrium. In hydra, a fresh water coelenterate, far
higher transepithelial rest potentials have been measured, from
10mV to 20mV, coelenteron also positive compared to the surrounding
water (Macklin and Westbrook 1976). The authors could show that
this transepithelial rest potential in hydra is highly dependent
upon osmo-and ionoregulatory processes. This regulatory capacity is
important for fresh water organisms but very much less so for marine
invertebrates. Transintegumentary uptake of DOM has so far never
been detected in any species of freshwater invertebrates (Stephens
and Schinske 1961). But there is still some speculation about the
transepithelial potential in hydra being related to either metabolic activity or transport mechanisms (Taddei-Ferretti and co-
workers 1976).
Changes in the value of the transepithelial potential due to mechanisms used for the uptake of DOM have been found for example in the
epithelium of the small intestine and kidney tubules of rats. In the
latter case rest potentials of +500 µV were reduced (!) to -300 µV
by adding 5 mM glucose (Frömter 1979). But it could be demonstrated
that the addition of glucose to the mucosal fluid increased the
potential difference across the epithelium of the tortoise small
intestine; this glucose induced potential could be reduced by
phlorizine (Wright 1966). In sea anemones phlorizine does not reduce
the glucose induced transepithelial transport potential. This result
is in contrast to uptake studies with radioactively labelled glucose
where a decrease of glucose absorption by phlorizine was measurable
(Schlichter 1975).
Comparing the results gained with tentacle tissue of sea anemones
with those of epithelia of small intestine or kidney tubules one must
take into consideration the following differences, which may be responsible for the disagreement. In sea anemones the transepithelial
potentials are measured across three layers of tissue, consisting of
two cellular layers (epidermis and gastrodermis) and one non cellular layer, the mesogloea, as opposed to the one layer transepithelial
potentials in vertebrate tissue (Fig. 2). Especially the mesogloea
may function as an electric insulator, this probably being the cause

for the suppressed electric response, because as mentioned in the introduction the absorptive process takes place through the epidermal cell surface.

Membrane Potentials

The values of membrane rest potentials, stable for more than 90 seconds are low in comparison to those e. g. measured in tubules of ratkidney; for sea anemones - 4 mV, in tubules up to - 70 mV (Frömter 1979). As already mentioned it is easy to obtain high membrane potentials from epidermal cells of sea anemones for short periods but it is difficult to get a stable recording. For further studies the method must be improved. The positive increase of membrane potentials of some millivolts during absorption of e. g. glucose is in accordance with studies on tubules of ratkidney (Frömter 1979) or epithelial cells in the small intestine of rats (Okada and co-workers 1977). As discussed for membrane rest potentials the membrane transport potential measured in the epidermal cells of sea anemones are also lower than in other tissues studied. Especially for the exchange of the incubation medium during the registration of membrane potentials new methods must be developed.

The electrical events, to be registered from tentacle tissue of sea anemones so far are schematically summarized in Fig. 9. The increase of transepithelial transport potentials is a slow process, whereas the alterations of membrane potentials occur immediately after DOM is added to the incubation medium.

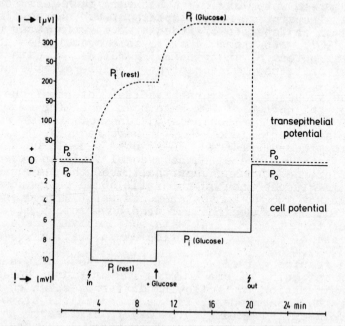

Fig. 9. Schematic diagram of the electrical events during the absorption of DOM by tentacle tissue of sea anemones.

Figure 10 shows in which way the electrical events during the penetration of a microelectrode through the tentacle wall of a sea anemone are related to components making up the tissue. The first negative going potential is due to the electrode tip coming through the first membrane into an epidermal cell after initial contact with mucous layer which apparently produces no electrical response. When the electrode then flexes through the mesogloea, potentials both positive and negative, can be recorded transiently. The last negative going potential is attributed to the electrode entering a gastrodermal cell and after the electrode has passed through the entire tentacle wall into the coelenteron a stable positive transepithelial rest potential can be recorded; see also Figs. 3, 4, 6. The first results of the investigations presented show that it is possible to study DOM uptake by sea anemones with electrophysiological methods. To gain more detailed information it is neccessary to modify the techniques used.

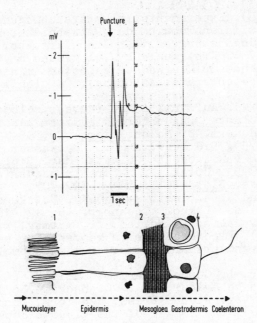

Fig. 10. Electrical events as related to the position of the microelectrode in the tentacle wall of Radianthus sp. The arrow indicates the direction of the perforation.

ACKNOWLEDGEMENT

These investigations were supported by the Deutsche Forschungsgemeinschaft. The authors are indepted to Miss H. Krisch for skilful technical assistance.

Frömter, E. (1979). Solute transport across epithelia: what can we learn from micropunkture studies on kidney tubules? J. Physiol., 288, 1-31.

Giordana, B. and V. F. Sacchi (1978). Glycine and L-alanine influence on transepithelial electrical potential difference in the midgut of Bombyx mori Larva in vitro. Comp. Biochem. Physiol., 61A, 605-609.

Gomme, J. (1979). D-glucose transport across the apical membrane of the surface epithelium in Nereis diversicolor. J. Membrane Biol. (in press).

Macklin, M. and G. Westbrook (1976). Ionic requirements of transepithelial potentials in isolated cell layers of hydra. In G. O. Mackie (Ed.), Coelenterate Ecology and Behavior, Plenum Press, New York, London. pp 715-727.

Okada, Y., W. Tsuchiya, A. Irimajiri and A. Inouye (1977). Electrical properties and active solute transport in rat small intestine. J. Membrane Biol., 31, 205-219.

Samarzija, I. (1978). Experimentelle Untersuchungen zum Aminosäurentransport im proximalen Tubulus der Rattenniere. Dissertation, Johann Wolfgang Goethe-Universität Frankfurt, pp. 1-96.

Schlichter, D. (1973). Ernährungsphysiologische und ökologische Aspekte der Aufnahme in Meerwasser gelöster Aminosäuren durch Anemonia sulcata (Coelenterata, Anthozoa). Oecologia (Berlin), 11, 315-350.

Schlichter, D. (1974). Aufnahme in Meerwasser gelöster Aminosäuren durch Anemonia sulcata: Das unterschiedliche Resorptionsvermögen von Ekto- und Endoderm. Z. Morph. Tiere, 29, 65-74.

Schlichter, D. (1975). Die Bedeutung in Meerwasser gelöster Glucose für die Ernährung von Anemonia sulcata. Mar. Biol., 29, 283-293.

Stephens, G. C. (1975). Uptake of naturally occurring primary amines by marine annelids. Biol. Bull., 149, 397-407.

Stephens, G. C., and R. A. Schinske (1961). Uptake of amino acids by marine invertebrates. Limnol. Oceanogr., 6, 175-181.

Undenfriend, S., S. Stein, P. Böhlen, W. Dairman, W. Leimgruber and M. Weigele (1972). Fluorescamine: a reagent for assay of amino acids, peptides, and primary amines in the picomole range. Science, 178, 871-872.

Taddei-Ferretti, C., L. Cordella and S. Chillemi (1976). Analysis of hydra contraction behaviour. In G. O. Mackie (Ed.), Coelenterate Ecology and Behavior, Plenum Press, New York, London. pp. 685-694.

Wright, E. M. (1966). The origin of the glucose dependent increase in the potential difference across the tortoise small intestine. J. Physiol., 185, 486-500.

THE ROLE OF NUTRITION IN THE ESTABLISHMENT OF THE GREEN HYDRA SYMBIOSIS

D. C. SMITH

Department of Botany, Bristol University, U.K.

ABSTRACT

Green hydra derives nutritional benefit from its Chlorella symbionts because they release substantial amounts of maltose to the animal tissue, which then rapidly metabolises it. Chlorella symbionts in culture do not release maltose, but regain this property after reinfecting their hosts. Maltose is not produced by free-living Chlorella, so maltose synthesis is uniquely characteristic of the symbiosis. Aposymbiotic hydra are moderately specific for their symbionts. There is little evidence that hydra recognises symbionts at first surface contact. Rather, establishment of the symbiosis is similar to a process of ecological colonisation, in which potential symbionts have to overcome a series of obstacles before reaching the far end of a host digestive cell. The key feature which then distinguishes symbiotic from non-symbiotic algae is that the former can divide but the latter cannot. The hypothesis is proposed that host cells permit symbiotic algae to divide because they 'recognise' their capacity or potential for maltose synthesis and release.

KEYWORDS

Symbiosis; Hydra viridissima; green hydra; Chlorella; phagocytosis; cell recognition; Coelenterate nutrition; maltose.

INTRODUCTION

The algal symbionts of the lower Metazoa are generally believed to be of nutritional benefit to their animal hosts. It is noteworthy that only a very few types of algae are able to establish a successful symbiosis with lower Metazoans (Smith, 1979). In freshwater associations, the symbionts almost always belong to the genus Chlorella, although many other genera of unicells occur in the habitat. The animal hosts are thus specific or highly specific as to their algal partners.

In green hydra, various attempts by American workers to establish a symbiosis with cultures of free-living Chlorella have so far been unsuccessful (Muscatine and co-workers, 1975; Hohman, unpublished). Indeed, consistent success in establishing a symbiosis was achieved only with algae freshly isolated from green hydra;

129

attempts to grow the symbionts of American hydra in pure culture have failed. The symbionts of European hydra have, however, been successfully cultured (Jolley & Smith, 1978).

This paper will examine the extent to which the specificity of the host for its symbiont depends upon the ability of the alga to provide nutrients. The paper will be divided into three parts:

a) A description of how the alga contributes to the nutrition of green hydra;
b) A description of some recent experiments on the establishment of the symbiosis in green hydra; and
c) A discussion of the possible importance of the nutritive mechanism of the alga in determining its specificity.

It is important in considering experimental work in green hydra to realise that different laboratory strains may show important morphological and physiological differences from each other in both animal and algal characteristics.

MECHANISMS BY WHICH ALGAL SYMBIONTS CONTRIBUTE TO THE NUTRITION OF THE ANIMAL HOST.

In parasitic and mutualistic symbiotic associations, there are two main types of nutrition: necrotrophy and biotrophy. Necrotrophy describes the situation in which the host first kills cells of the symbiont, and then digests the dead products. Biotrophy describes the situation most commonly found in mutualistic symbiosis, in which the symbionts stay alive, and substantial amounts of nutrients continuously pass from the living cells of the symbiont to the animal host. Biotrophy is a more advanced form of nutrition than necrotrophy, because the host makes continued use of the biochemical properties of the symbiont, instead of destroying them. The success of biotrophy does, however, depend upon the symbiont being induced to release nutrients to the host.

In green hydra, there is no doubt that the nutrition is biotrophic since it is very rare to see algal cells being digested. Therefore, a situation exists in which living algal cells are continuously releasing nutrients to the host. The quantitative importance of this flow of nutrients, especially under conditions of low food supply, was illustrated by the original experiments of Muscatine and Lenhoff (1965). They showed that under conditions of reduced food supply, green hydra grew much better than albino hydra. Since the faster growth of green hydra only occurs if they are in the light, then it is likely that it is products of photosynthesis which are being passed to the host. This was further investigated by Lenhoff and Muscatine (1963), who incubated animals in the light in solutions of $NaH^{14}CO_3$, and traced the movement of photosynthetically fixed carbon from endoderm to ectoderm. A more accurate picture of the amount of transfer from alga to animal is obtained if animals are homogenised, and the algal cells separated by centrifugation. Such experiments show that as much as 30%-40% of the ^{14}C fixed by algal photosynthesis moves rapidly to the animal tissue in the 'Florida' strain of green hydra (Eisenstadt, 1971). The kind of compound released from the algae has been studied by allowing algae to fix $^{14}CO_2$ immediately after isolation from the animal. Although many compounds become rapidly labelled with ^{14}C within the algal cell, only one product is released in any quantity to the medium, the disaccharide maltose. The maltose that is released to the medium is probably synthesised on the surface of the algal cells (Cernichiari, Muscatine & Smith, 1968) since (a) maltose cannot be detected inside the cells; and (b) maltose synthesis is very sensitive to the pH of the medium, as if the enzyme(s) involved were on the cell surface.

In the case of 'European' green hydra, Jolley and Smith (1978) were able to isolate
the alga into culture. The cultured symbiont does not synthesise and release
maltose (indeed, no free-living Chlorella can do this), but when reintroduced into
hydra, it developed this property again. Thus, the synthesis and release of large
amounts of maltose is a characteristic which specifically develops in the symbiosis.

To demonstrate the nutritional value of the symbiont, it is necessary to show that
the animal can metabolise the maltose it receives from the alga. Mews (unpublished)
showed that animal tissues contained enzymes capable of rapid breakdown of maltose.
She gave a brief 10 minute pulse of $^{14}CO_2$ to 'European' green hydra, and then
investigated the fate of the ^{14}C in the animal tissue. Her results (Fig. 1)
clearly show that the ^{14}C received from the alga is metabolised.

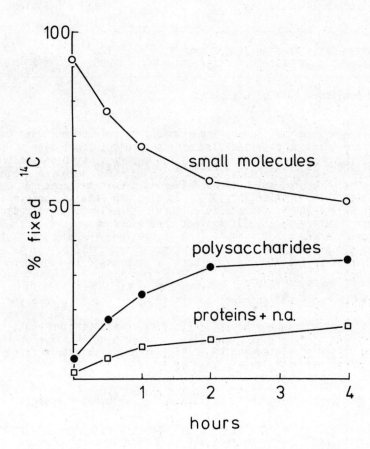

Fig. 1.   Metabolism of a pulse of fixed ^{14}C by the animal
          tissues of green hydra. Animals exposed to $NaH^{14}CO_3$
          in the light for 10 minutes, then rapidly washed and
          transferred to non-radioactive media; at intervals,
          animals homogenised, algae removed by centrifugation,
          and ^{14}C distribution determined in remaining animal
          fraction. n.a. = nucleic acids. Data of L.K. Mews,
          unpublished.

Much of the fixed ^{14}C in the animal tissue accumulated as polysaccharides, mostly the reserve carbohydrate, glycogen. She compared green and albino hydra, and found that the former contained more maltase enzyme and more glycogen than the latter. The importance of glycogen as a reserve can be noted if animals are starved. In both green and albino, glycogen becomes depleted, but the larger glycogen reserve of green hydra suggest it could withstand this aspect of starvation for a much longer period (Table 1).

TABLE 1. Effect of starvation on glycogen levels in 'European' hydra.

Glycogen level (μg glucose/mg protein)	Green	Albino
After daily feeding for 9 days	128	62
After 9 days subsequent starvation	96	23

Unpublished data of L.K. Mews.

Not only does the host animal induce the symbiont to release maltose, but it also exerts a strict regulation over the size of its symbiont population. McAuley (1979) has found that there is remarkable constancy in the ratio $\frac{symbiont\ volume}{animal\ protein}$ between different strains of hydra, and between different growth temperatures. Even under optimum conditions, Jolley & Smith (1978) found growth rate of the symbiont in green hydra is only 1/20 of its rate in culture: clearly the host restricts and controls growth. Interestingly, if ambient conditions change, then the ratio of symbiont to host changes. For example, the population of symbionts declines sharply if animals are kept in constant darkness (Pardy, 1974). It is as if, when the symbiont cannot produce food, its growth is suppressed to reduce the burden of demand on the host.

In summary, the mechanism of nutrient supply from alga to host involves:

a) the induction of a unique process, maltose synthesis, in the alga; and
b) the development of strict and sensitive controls over the size of the symbiont population, clearly related to the ability of the symbiont to supply food, and the need of the host for it.

BIOLOGY OF THE ESTABLISHMENT OF THE GREEN HYDRA SYMBIOSIS

The system for studying the establishment of the symbiosis in green hydra is relatively simple, and depends on the fact that there are a number of simple techniques by which hydra can be freed from its symbionts rapidly and in large numbers (e.g. Pardy, 1976). The symbiosis can then be re-established by introducing a suspension of algae into the coelenteron by micro injection through the mouth. If freshly isolated symbionts are used, the following sequence of events, first observed by Pardy and Muscatine (1973), occurs. After the algae come into contact with the external surface of the digestive cells, they become rapidly engulfed and enter the cell by phagocytosis. The algae remain in individual vacuoles, and then migrate over a five hour period to the far end of the cell. Over the next few weeks they divide and grow until the normal population of algae is reached. If unacceptable algae are used, they enter, but soon are aggregated into large vacuoles and expelled within about 5 hours, instead of remaining in individual vacuoles and migrating.

The general problem of how one cell recognises another is currently being widely studied. Nearly all investigations have been concerned with recognition between cells of the same species in problems such as embryo development, cell adhesion, tissue culture, fusion between sperm and egg, pollen and style, etc. These investigations lead to a general conclusion on the part of many authors that cells recognise each other by the interaction of certain kinds of molecules on the cell surfaces. In particular, glycoproteins are often considered to be very important in such 'recognition' interactions. Compounds which bind to specific sugar moieties on glycoproteins will also block recognition - and lectins such as Concanavalin A are amongst the best known of these. However, the recognition of an alga by a digestive cell of hydra is recognition between two unlike or genetically very dissimilar cells, a problem much less studied in biology. Unfortunately, many of the theories and techniques applied to recognition of 'like-for-like' are applied rather uncritically to recognition between unlike cells. Recently, a number of papers have appeared presenting evidence which, by analogy with 'like-for-like' studies, could be interpreted as indicating that recognition of potential symbionts occurs at first surface contact. There are three types of such evidence: (i) Pool (personal communication), later confirmed by McNeil (1980), showed that heat-killed symbionts and latex spheres were taken up more slowly than live symbionts by digestive cells; (ii) Pool (1979) described experiments suggesting that antigenic determinants localised on the algal cell surface were responsible for the phagocytotic recognition of potential symbionts; and (iii) Meints and Pardy (unpublished) and Pool (unpublished) find that use of Concanavalin A under various conditions will either reduce or even totally prevent uptake of live symbionts. However, such evidence presents the following problems. (a) The experiments only concern the first 6 to 24 hours after infection - they do not continue for the 2-4 weeks necessary to show that a permanent association, with a normal population of algae, has been formed. (b) Symbionts are always isolated from homogenates of live hydra, so the problem of contamination of algal cell surface by animal material always remains. McNeil (1980) found his isolated algae to be free of contaminating particles when examined by electron microscopy, but Mews (unpublished) found traces of animal protein still present after repeated washing of her algae. (c) No conclusive controls were included in the experiments to show that Concanavalin A specifically affects only uptake of symbionts, and not other kinds of algae. (d) There is no proof that a high rate of phagocytosis implies symbiont recognition as distinct from some factor concerned only with the act of phagocytosis itself.

Reinfection experiments with 'European' hydra

Recently, Jolley & Smith (1980) investigated the biology of reinfection in 'European' hydra, which has the advantage that its algal symbiont can be cultured (i.e. so as to eliminate any animal contamination). Initially, they observed the same sequence of events in the early stages of uptake as that described by Pardy and Muscatine for American strains. However, when the progress of reinfection over a prolonged period was studied, a period of decline in numbers of algae occurred after the initial infection, before re-population began (Fig. 2). This phase was termed 'resorting'; its cause was not known, but it showed there is a period when the population of algae declines to a very low level. When Jolley and Smith infected 'European' hydra with a range of different algae - and infected a sufficient number of animals to overcome this fierce 'resorting' process, the pattern of specificity shown in Table 2 was found. Clearly 'European' hydra is much less specific than was reported for 'Florida' by Muscatine and co-workers (1975). In 'European' hydra there is a group of algae which are intermediate in their capacity to infect - they can persist in digestive cells for very long

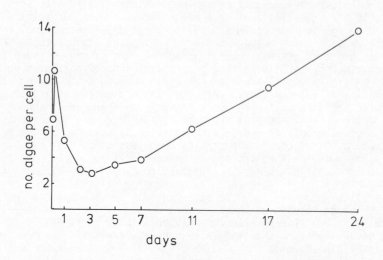

Fig. 2.   Progress of infection of aposymbiotic 'European'
           green hydra by freshly isolated 'European' algae.
           Ordinate gives the mean number of algae per infected
           digestive cell. Data of Jolley and Smith, 1980.

periods of time (up to 14-21 days) before ejection, but they never divide.   Jolley
and Smith (1980) found that 'Florida' hydra in fact showed a similar picture, but
were a little more difficult to infect;   'resorting' was less distinct in 'Florida'

TABLE 2.   Specificity of 'European' green hydra: summary of
            reinfection experiments.

Successful infection Alga	Sporo-pollenin	Mesotrophic	Originally symbiotic
'European' (newly isolated)	+	.	+
'Florida' (newly isolated)	+	.	+
'European' (cultured)	+	+	+
NC64A (Muscatine)	+	+	+ (Paramecium)
C. protothecoides 211/6	-	+	+ (Paramecium)
C. protothecoides 211/7a	-	+	- (tree-sap)
Unsuccessful infection			
NC64A (Pardy)	+	+?	+ (Paramecium)
C. vulgaris 211/11b	+	-	-
C. vulgaris 211/1e	-	-	-

Table continued on following page

TABLE 2. (Continued) Specificity of 'European' green hydra:
summary of reinfection experiments.

Alga	Sporo-pollenin	Mesotrophic	Originally symbiotic
Partial success			
Chlorella from R. Frome	?	-	- (hydra habitat)
C.fusca var. vacuolata 211/8b	+	-	-
Oocystis	?	?	-

Data of Jolley and Smith (1980).

Jolley and Smith (1980) compared uptake rates of different kinds of algae and found little correlation between initial rate of uptake and eventual capacity to establish a symbiosis. If symbionts were damaged only gently - i.e. heating to 60° for 10 min, not 100° for 5 min - then they entered at a higher rate than undamaged symbionts. Latex spheres of similar diameter to symbionts also entered at a higher rate. Indeed, the rates of uptake of latex spheres can be appreciably increased by including glycogen in the suspending medium (Table 3). Thus, it was concluded that the rate of uptake has little to do with whether or not the object is a symbiont.

TABLE 3. Effect of 1% glycogen on uptake of latex spheres
in 'European' hydra.

	- glycogen	+ glycogen
% digestive cells with spheres	51%	78%
Mean no. spheres per infected cell	5.5	13.7

Results 2 hours after injection of spheres into 2-day
starved aposymbionts. Data of Jolley and Smith (1980).

It was reasonably concluded from Pardy and Muscatine's (1973) studies of reinfection that an important stage in specific recognition occurred after phagocytosis, in that symbionts remained within individual vacuoles and migrated to the distal end, while non-symbionts were aggregated into large vacuoles and ejected. However, Jolley and Smith (1980) found that if latex beads are suspended in 0.1% Bovine Serum Albumin (B.S.A.) and injected into starved hydra, they will also stay in individual vacuoles and migrate (Table 4). This makes it less easy to believe that migration is the result of specific recognition. Since Burnett (1959) observed glycogen particles to migrate at a rate which we now know corresponds to the rate for symbiotic algae, it could be that the effect of B.S.A. on latex particles is to

TABLE 4.   Effect of 0.1% Bovine Serum Albumen (BSA) on
           uptake and migration of latex spheres (3.9 μ
           diameter) in 'European' hydra.

	- BSA	+ BSA
% digestive cells with spheres	12%	53%
Mean no. spheres per 'infected' cell	3.5	3.6
% spheres migrating to distal end	3%	56%

Results 24 h after injection of spheres into 4-day starved
aposymbionts. Data of Jolley and Smith, 1980.

cause them to mimic food particles. Alternatively, B.S.A. could be affecting the
surface charge on the latex spheres. Whatever the explanation, the fact that some
non-symbiotic algae can migrate also indicates that it is not the main 'recognition'
event. Jolley and Smith (1980) considered that the arrival of an alga at the
distal end of a digestive cell is not a consequence of a positive and specific act
of recognition. Rather, it is the result of successfully overcoming a series of
obstacles: gaining entry to the cell, resisting or avoiding host digestive
enzymes, avoiding aggregation into large vacuoles and then ejection, and
successfully mimicking the right qualities needed to participate in the cellular
migration system. Thus, for Jolley and Smith, reaching the distal end of a
vegetative cell is more akin to ecological colonisation than processes associated
with cell recognition in 'like-for-like' situations. The stages of reinfection of
Jolley and Smith are compared with those of Muscatine and Pardy in Table 5.

TABLE 5.   Stages in the reinfection of green hydra

Muscatine and co-workers (1975	Jolley & Smith (1980) for 'European' hydra
CONTACT	CONTACT
ENGULFMENT	ENGULFMENT
RECOGNITION	
MIGRATION	MIGRATION
	RESORTING
REPOPULATION	REPOPULATION

The only quality which absolutely distinguishes symbionts from non-symbionts is
that symbionts can divide, while non-symbionts do not. From evidence given above,
it is clear that the host regulates the growth of the symbiont. Also, there are
no reports of algal parasites of hydra which can grow uncontrolled. The inference
from this is that the host 'recognises' a quality in the alga, and then 'permits'
it to divide. There is no direct evidence, but by far the most likely 'quality'
that a host would recognise would be the release of nutrients. Thus, it becomes

a reasonable hypothesis that specific symbiont recognition in green hydra is
concerned with nutrition.

POSSIBLE IMPORTANCE OF NUTRIENT RELEASE IN DETERMINING ALGAL SPECIFICITY

The green hydra symbiosis presents a clear cut situation when the alga inside a
cell synthesises and releases maltose, while the alga in culture does not. There
are three possibilities by which maltose synthesis could be induced:

Modified glycosyl transferase. This hypothesis was first advanced by Jolley and
Smith (1978). They suggested that the animal produces a protein which binds to
surface glycosyl transferases on the alga, so changing their specificity that they
bind glucose to glucose instead of to proteins, thus producing maltose. An
analogy was drawn to the induction of lactose synthesis in mammary gland cells.
At the onset of lactation, the protein α-lactalbumin is secreted and binds to a
galactosyl transferase enzyme, changing the specificity of the enzyme such that it
binds galactose to glucose (instead of to glycoproteins) so producing lactose.
Since α-lactalbumin has an amino-acid sequence indistinguishable from the common
digestive enzyme, lysozyme, it is possible to construct an attractive theory for a
comparable protein to be produced by hydra. Such a protein could be one of the
digestive enzymes secreted into the phagocytotic vacuole containing the alga. Its
interaction with algal surface enzymes could make a logical basis for symbiont
specificity. It would be a very direct way in which a host 'recognises' the ability
of an alga to provide nutrients. It should be stressed that so far we lack any
direct evidence for this otherwise attractive theory.

Maltose production during phosphate sequestration. This idea derives from
experiments on sugar beet leaves by Herold, McGee and Lewis (1980). Normally,
there is no detectable maltose in these leaves, but if the sugar mannose is added,
then substantial amounts of maltose are produced. This is because when mannose
enters cells, it is rapidly phosphorylated but not further metabolised. All the
free inorganic phosphate in the cytoplasm becomes immobilised as non-metabolisable
mannose phosphate. This produces a situation where reactions are favoured which
generate inorganic phosphate from other sugar phosphates. For example, a
phosphorylase enzyme might act: glucose -1- phosphate + glucose = maltose +
inorganic phosphate. In the context of green hydra, such a situation might arise
if the host animal could remove inorganic phosphate from proximity to the surface
enzymes of the symbionts, so generating conditions favouring maltose synthesis.
Although this offers a very satisfactory solution to the problem of maltose
production in symbiosis, it is less easy to explain how it causes high specificity
on the part of the host, unless it be due to a specific system of phosphate
sequestration.

Adherence of animal maltase to algal cell surface. Since, theoretically, all
enzymes can work in reverse, the enzyme which splits maltose ought, under certain
conditions, to catalyse maltose synthesis. Since Mews (unpublished) could not
remove all the animal protein from a suspension of freshly isolated algae by
repeated washings, and since she also showed that meso-erythritol can inhibit
both animal maltase activity and algal maltose synthesis, it is possible that
maltose is in fact produced by an animal enzyme intimately adhering to the algal
cell surface. Specificity is then determined by those algae which can create
the right microenvironment for the animal enzyme to work in reverse.

CONCLUSIONS

If the key advantage to a host is nutrient release from a symbiont, it is logical
to assume that recognition of nutrient release should be a key factor in determining specificity, especially when the nutrient released is only produced in
symbiosis.

In so far as green hydra is concerned, it is felt that knowledge of how maltose
synthesis is induced would be particularly helpful to the understanding of what
determines specificity between host and symbiont. This is not to deny the value
of 'conventional' recognition studies of surface glycoproteins, etc., but it must
be stressed that alternative mechanisms may well be involved where hosts are
recognising organisms genetically very different from themselves.

ACKNOWLEDGEMENTS

I am grateful to the Science Research Council and the North Atlantic Treaty
Organisation for financial support. I am grateful to R. Pool, T. Hohman, R. Pardy,
P. McNeil for comments and for letting me see unpublished manuscripts. I am
especially indebted to Miss L.K. Mews for permission to quote from unpublished
results.

REFERENCES

Burnett, A. (1959). Histophysiology of growth in hydra. J. exp. Zoology, 140,
        281-334.
Cernichiari, E., L. Muscatine and D.C. Smith (1969). Maltose excretion by the
        symbiotic algae of Hydra viridis. Proc. R. Soc. B, 173, 557-576.
Eisenstadt, E. (1971). Transfer of photosynthetic products from symbiotic algae
        to animal tissues in Chlorohydra viridissima. In H.M. Lenhoff, L. Muscatine
        and L.V. Davies (Eds.), Experimental Coelenterate Biology, U. Hawaii Press,
        Honolulu. pp. 202-208.
Herold, A., E.E.M. McGee and D.H. Lewis (1980). The effect of orthophosphate
        concentration and exogenously applied sugars on the distribution of newly
        fixed carbon on sugar beet leaf discs. New Phytol., 85, in the press.
Jolley, E., and D.C. Smith (1978). The green hydra symbiosis. I. Isolation, culture
        and characteristics of the Chlorella symbiont of 'European' Hydra viridis.
        New Phytol., 81, 637-645.
Jolley, E., and D.C. Smith (1980). The green hydra symbiosis. II. The biology of
        the establishment of the association. Proc. R. Soc. B, in the press.
Lenhoff, H.M., and L. Muscatine (1963). Symbiosis: on the role of algae symbiotic
        with Hydra. Science, 142, 956-958.
McAuley, P. (1979). Studies of the green hydra symbiosis. Ph.D. thesis, University
        of Bristol.
McNeil, P. (1980). Mechanisms of nutritive endocytosis. I. Phagocytotic versatility
        and cellular recognition in the Chlorohydra digestive cell, an SEM study.
        J. Cell Biol., in the press.
Muscatine, L., C.B. Cook, R.L. Pardy and R.R. Pool (1975). Uptake, recognition
        and maintenance of symbiotic Chlorella by Hydra viridis. Symp. Soc. exp. Biol.,
        29, 175-203.
Muscatine, L., and H.M. Lenhoff (1965). Symbiosis of hydra and algae. I. Effects
        of some environmental cations on growth of symbiotic and aposymbiotic hydra.
        Biol. Bull., 219, 316-328.

Pardy, R.L. (1974). Some factors affecting the growth and distribution of the
    algal endosymbionts of Hydra viridis. Biol. Bull., 147, 105-118.
Pardy, R.L. (1976). The production of aposymbiotic hydra by the photodestruction
    of green hydra zoochlorellae. Biol. Bull., 151, 225-235.
Pardy, R.L., and L. Muscatine (1973). Recognition of symbiotic algae by Hydra
    viridis. A quantitative study of the uptake of living algae by aposymbiotic
    H. viridis. Biol. Bull., 145, 565-579.
Pool, R.R. (1979). The role of algal antigenic determinants in the recognition of
    potential algal symbionts by cells of Chlorohydra. J. Cell Sci., 35, 367-379.
Smith, D.C. (1979). From extracellular to intracellular: the establishment of a
    symbiosis. Proc. R. Soc. B, 204, 115-130.

DIGESTION ET NUTRITION CHEZ LES ACTINIAIRES ET LES CERIANTHAIRES

Y. TIFFON* ET M. VAN PRAËT**

*Université de Caen, 14032 Caen Cedex, France
**Muséum National d'Histoires Naturelles, Laboratoire de Biologie des Invertébrés Marins, 57 Rue Cuvier, 75005 Paris, France

ABSTRACT

Analysis of the feeding response of the Cerianthariae and Actiniariae indicates two phases : preparatory activity and ingestion. The chemical activators of these two phases are usually aminoacids and proteins. Extra cellular digestion occurs in the gastroderm. Proteases have been detected by histochemical and electron microscopical techniques in the filaments. Intake of particulate matter occurs in the endoderm of the filaments. Hydrolases active in the acid range were characterised in homogenates of sterile septae. Acid phosphatase is present in lysosomes and in newly formed phagosomes. Dissolved amino acids, carbohydrates and proteins are utilised by Cerianthae and Actiniae. Uptake takes place in the endoderm but also in the ectoderm of the column and of the tentacules. There are few informations concerning the excretory products of protein activity. Cerianthae do not harbour zooxanthellae. Symbiots are confined to the endodermis of Actiniae. They are capable of active photosynthesis and appreciable amount of photosynthetically fixed carbon pass to the animal tissues. Among the compounds release, glycerol is particularly important. The translocated products are mainly incorporated in lipids and proteins of the host. Digestion of algae has been reported in one association.

KEYWORDS

Feeding response ; proteases ; acid hydrolases ; lysosomes ; uptake ; dissolved organic matter ; zooxanthellae ; released fixed carbon.

INTRODUCTION

Les Actinaires et les Cérianthaires paraissent avoir résolu le problème de leur nutrition avec beaucoup de succès. La plupart des données sur la nutrition des Actinaires et des Cérianthaires proviennent des résultats expérimentaux ou d'observations réalisées sur des animaux en élevage. En circuit ouvert mais en l'absence de proies macroscopiques, les Cérianthaires paraissent pouvoir vivre indéfiniment. Leur aptitude à collecter des substances organiques en suspension ou en solution paraît donc remarquable. Des Anemonia sulcata en aquarium, rendues aposymbiotiques vivent très facilement à l'obscurité en acceptant des proies macroscopiques. Des Metridium pourtant réputées microphages sont nourries exclusivement en circuit fermé avec des proies macroscopiques. Ainsi les Cérianthaires comme les Actiniaires sont

capables de beaucoup d'opportunisme et leur succès dans les biotopes où on les
rencontre est certainement dû au fait que leurs mécanismes de nutrition sont d'une
remarquable efficacité.

CAPTURE DE LA NOURRITURE ET COMPORTEMENT DE CAPTURE

En présence d'une proie expérimentale, un Cérianthe ou une Actinie s'épanouit et ses
tentacules s'agitent. Chez Metridium senile (Pantin, 1950) il y a de plus expansion
du disque oral et balancement de la colonne. Cette étape de précapture peut être
induite par la seule présence dans l'aquarium de substances nutritives et relève
donc d'une chémoréception. Outre le stimulus chimique, la phase de préhension nécessite
un contact. Lindstedt, Muscatine et Lenhoff (1968) ont découvert que la valine à la
concentration de 10^{-5}M permet de faire ingérer des particules inertes normalement
refusées. Needler Arai et Valder (1973) ont analysé le comportement alimentaire de
Pachycerianthus fimbriatus. La réponse comprend deux phases : transfert de la proie
après contact avec les tentacules marginaux aux tentacules labiaux et à la bouche
puis ingestion. Le transfert est activé par une action mécanique directe ou par la
stimulation des tentacules labiaux par des homogénats de proie. L'arginine agit comme
un activateur de l'ingestion. Lindstedt (1971) a montré chez Anthopleura que l'aspa-
ragine contrôle la contraction des tentacules et le transfert des proies à la bouche.
Le glutathion réduit contrôle l'ingestion. La plupart des travaux concernent la
stimulation de la nutrition des Actiniaires ont utilisé des acides aminés, mais
Williams (1972) a montré que des vitamines du groupe B avaient aussi une action.
Les chémorécepteurs seraient ectodermiques et placés sur la colonne et le disque
oral. Il n'a cependant pas été fourni une preuve formelle de la conduction de
l'information (Mac Farlane, 1970). Un jeûne prolongé abaisse le seuil de réponse, si
bien que la décharge des cnidocytes qui interviennent dans la capture d'une part et
le mouvement d'ingestion d'autre part peuvent être induits par une stimulation mécanique seule.

DIGESTION EXTRACELLULAIRE

La digestion chez les Actiniaires a donné lieu à de nombreuses controverses. Mesnil
(1901), élève de Metschnikoff considérait que seule la digestion intracellulaire
existait chez les Actiniaires. Mais l'idée d'une digestion effectuée en deux temps
s'est rapidement imposée. Krukenberg (1878) avait déjà pu extraire par la glycérine
un ferment trypsique à partir d'un broyat de gastroderme d'Actinie. Les frères
Hertwig (1880) attribuaient à certaines cellules glandulaires des filaments une
activité dans l'élaboration des sucs digestifs. Toutefois, les travaux anciens tels
ceux de Chapeaux (1893) ou de Jordan (1907) se sont souvent bornés à la préparation
d'extraits bruts de gastroderme et les conclusions obtenues n'établissent pas la
différence entre enzymes d'origine extracellulaire ou enzymes d'origine intracellulaire
Krijgsam et Talbot (1953), Nicol (1959) reprennent l'étude de la digestion des protéines et confirment chez Calliactis et Tealia le manque d'activité des liquides
Coelentériques contrastant avec la très grande activité des broyats. Ils concluent
à une digestion de contact. Jeuniaux (1962) nourrit des Adamsia, des Anemonia et des
Anthopleura avec des fragments de chitine (apodèmes de Maia décalcifiés). La mise en
évidence d'une digestion extra-cellulaire à partir de fragments de taille relativement
élevée joint à l'observation d'une altération inégale des fragments de chitine et à
la présence de nombreux filaments mésentériques à la surface des fragments conduit
à penser que la digestion de la chitine se déroule selon un procédé de contact.
Gibson et Dixon (1969) ont isolé deux protéases A et B des tissus glandulaires de
Metridium. L'activité de ces deux enzymes est révélée en utilisant l'ATEE comme
substrat. L'activité protéolytique de chacune de ces enzymes est abolie par le DFP
et le TPCK. Des études de spécificité et d'inhibition, les auteurs concluent que la
protéase A ressemble à la chymotrypsine des mammifères. La présence possible d'un

précurseur inactif de la chymotrypsine a été mis en évidence. Chez <u>Cerianthus lloydi</u>
(Tiffon et Bouillon, 1975) ont mis en évidence une trypsine, une chymotrypsine et un
trypsinogène.

Ces enzymes sont contenues dans des cellules glandulaires situées dans le bourrelet
glandulaire des entéroïdes. Bien que n'ayant pas la même morphologie chez les
Cérianthaires et les Actiniaires où il a été décrit, ce bourrelet glandulaire
renferme outre les cnidocytes des cellules glandulaires identiques. Elles comprennent
des cellules à mucus de deux catégories et des cellules à grains de zymogène
p-dimethylaminobenzaldehyde positives. Ces grains de zymogène sont également présents
dans le pharynx (Vader et Lönning, 1975 ; Tiffon et Bouillon, 1975 ; Van Praët, 1978).
La morphologie et le mode d'élaboration de ces grains de zymogène paraissent être
identiques à ceux décrits pour le pancréas des mammifères. Ceci avait d'ailleurs
déjà été observé par Jamieson dans le cas de <u>Metridium</u> (Gibson et Dixon, 1969).
Faute de techniques appropriées, les carboxypeptidases, les aminopeptidases et les
dipeptidases n'ont pas été recherchées dans ces structures. Des expériences avec
des éléments marqués ont toutefois permis de démontrer que les acides aminés en
provenance de la proie sont incorporés aux protéines deux heures après l'ingestion
ce qui semble prouver que les protéines sont hydrolysées jusqu'au stade acide aminé
dans la cavité gastrique (Murdock, 1971).

ABSORPTION DES ACIDES AMINES ET DES SUCRES EN SOLUTION

L'absorption des substances dissoutes est une capacité générale des animaux marins
(Jørgensen, 1976). Chez les Actiniaires et les Cérianthaires, les auteurs sont signalé
des incorporations rapides d'acides aminés et de sucres, surtout dans l'ectoderme
qui paraît être le site privilégié d'absorption de ces produits (Chia, 1972
Schlichter, 1974,1975 ; Tiffon et Daireaux, 1974 ; Van Praët, 1978). Mais lorsque
les cloisons sont excisées, et mises en contact avec une solution renfermant un
hydrolysat de chlorelles, une incorporation est signalée au niveau des tractus glandulaires ciliés des entéroïdes. L'incorporation et donc vraisemblablement l'absorption
est particulièrement importante dans l'ectoderme des tentacules. Encore ici dans le
cas des Cérianthes l'ectoderme des tentacules labiaux n'ont rien incorporé malgré un
contact prolongé jusqu'à quarante heures (Tiffon et Daireaux, 1974). Selon Schlichter
(1978), des phénomènes d'inhibition compétitifs existent entre les acides aminés de
même charge. Il existe au moins trois systèmes de transport actif : un pour les
acides aminés basiques, un pour les acides aminés acides et un système à large
spécificité pour les neutres. L'absorption active peut s'exercer contre des gradients
très importants. Le glucose peut être 10^6 fois plus concentré dans les tissus que
dans l'eau de mer. Schlichter (1975) considère que pour <u>Anemonia sulcata</u>, 50 % du
métabolisme basal est assuré par l'absorption directe. Selon ce même auteur, certains
éléments sont immédiatement métabolisés tels le glucose, la sérine et la glycine.
Six heures après avoir placé des <u>Anemonia sulcata</u> dans un aquarium contenant du
glucose ^{3}H, la concentration en sucre est tombée de 100 µg à 20 µg mais la radio
activité totale est revenue à une valeur identique à celle du début.

PHAGOCYTOSE ET PINOCYTOSE

Les Coelentérés partagent avec d'autres invertébrés inférieurs une très grande
aptitude à la digestion intracellulaire. Afzelius et Rosen (1965) ont parlé à leur
propos de phagocytose digestive. Les travaux de Metschnikoff (1880) sur <u>Sagartia</u>
et <u>Aiptasia</u> et de son élève Mesnil (1901) devaient conduire à la notion d'actinio-
diastase. Cette vue un peu enthousiaste d'une enzyme universelle extra et intra -
cellulaire a été revue depuis, mais la notion d'une acidité du phagosome dans les
premiers stades de l'endocytose est demeurée. Des techniques cytochimiques ultra-
structurales ont permis de mettre en évidence des phosphatases acides dans les
cellules des cloisons septales des Cérianthes et des Actinies (Van Praët, 1976 ;

Tiffon et Hugon, 1977). Chez les animaux à jeûn, l'activité de la phosphatase acide
a été observée dans les citernes de l'appareil de Golgi, dans les lysosomes. Douze
heures après la prise d'un repas expérimental, une phosphatase acide est mise en
évidence dans les phagosomes. Huit jours après ce même repas, la phosphatase alcaline
est détectée dans des corps résiduels provenant des vacuoles digestives. Dans des
homogénats de gastroderme des Cérianthes, une β-glycérophosphatase, une β-glucuronidase
une cathepsine et une ribonucléase ont été caractérisées dont le maxima d'activité
est acide. Le Triton X 100, la congélation décongélation, une augmentation de la
température causent un accroissement de l'activité libre des hydrolases acides de
l'homogénat. D'autres caractères de sédimentabilité et de solubilisation montrent
que ces hydrolases sont contenues dans des lysosomes (Tiffon, 1973). Toutes les
cellules de l'endoderme paraissent capables de phagocytose. Seul l'encombrement des
proies paraît constituer une limitation au phénomène. C'est ainsi que des fibres
musculaires ne sont pas phagocytées par l'endoderme des tentacules des Cérianthes
tandis que des hématies de poulet le sont (Tiffon et Daireaux, 1974). Les possibilités
de pinocytose de protéines en solution ont été expérimentées au moyen de ferritine.
Une pinocytose par l'ectoderme a été montrée dans les tentacules labiaux de Cerianthus
lloydi (Tiffon et Daireaux, 1974) et par les acrorhages d'Actinia equina (Van Praët,
1978). Dans l'endoderme, l'intensité du phénomène varie avec les tissus. Elle est
intense pour la ferritine dans les cellules à vacuole et les cellules à concrétion
d'Actinia equina. Dans l'endoderme des Cérianthes, elle est plus forte dans l'endoderme
des cloisons fertiles que dans l'endoderme des cloisons stériles. Chez Actinia equina,
les lipides injectés dans la cavité entérique sont absorbés par les cellules phagocytaires et par certaines cellules des entéroïdes. Si l'expérience est prolongée
jusqu'à neuf jours, les lipides s'accumulent à la base des cellules phagocytaires
dans des inclusions qui peuvent envahir tout le cytoplasme de ces cellules. Le
tractus intermédiaire des entéroïdes et la région des mésentéroïdes prennent alors
l'aspect de tissu adipeux (Van Praët, 1978).

EXCRETION

Les éléments indigestibles résiduels demeurent parfois très longtemps à l'état de
post phagosomes. Mouchet (1929) rapporte une rétention de sept mois pour du fer dans
les cloisons d'Actinia equina. L'excrétion azotée n'a été l'objet que d'hypothèses.
On a cependant signalé dans les cellules du pharynx et des tractus réticulés de la
portion supérieure des entéroïdes (Mouchet, 1929 ; Van Praët, 1977) des grappes de
concrétions biréfringentes contenues dans des cellules. Ces cellules sont limitées
par une double membrane et la post-fixation osmiée n'y laisse persister qu'une trame
rayonnée.

SYMBIOSE CHEZ LES ACTINIES

Les Cérianthaires ne contiennent pas de zooxanthelles. Les zooxanthelles des
Actiniaires sont à rapporter au genre Gymnodinium adriaticum. La structure et le
cycle de ce Dinoflagellé ont été souvent décrits (Droop, 1963 ; Mc Laughlin et Zahl,
1966 ; Taylor, 1973). Notre préoccupation est de connaître quels sont les rapports
de ce symbiote avec l'hôte et d'apprécier son rôle dans l'économie de la nutrition
des actiniaires.

LOCALISATION DES ZOOXANTHELLES DANS LES TISSUS

Les symbiotes se trouvent toujours dans l'endoderme et plus particulièrement dans
l'endoderme des tentacules, du disque oral et des tractus à zooxanthelles de la
portion supérieure des cloisons. Il existe certains organes spécialisés qui contiennent
de très grandes quantités de symbiotes. C'est le cas des collerettes de Phyllactis

flosculifera. Ces collerettes sont épanouies pendant le jour et les tentacules sont
rétractés. C'est l'inverse pendant la nuit. Tout se passe donc comme si ce comportement devait favoriser la photosynthèse (Steele et Goreau, 1977).

REGULATION DE LA QUANTITE DE ZOOXANTHELLES DANS LES TISSUS

Steele (1976) a fourni un schéma de la régulation du nombre des symbiotes pour
Aiptasia tagetes. Le taux de multiplication des xanthelles paraît lié à l'illumination.
En même temps, leur nombre est maintenu constant par l'hôte qui utilise les xanthelles
en excès pour coloniser des tissus néoformés ou qui les expulse inclus dans des bou-
lettes de mucus. Dans les cas défavorables : vieillissement ou concentration excessive,
les algues produiraient des métabolites altérés qui induiraient une augmentation des
processus d'expulsion. Dans une espèce Phyllactis flosculifera, un processus d'utilisation alimentaire direct des algues a pu être mis en évidence (Steele et Goreau, 1977).
Cette espèce rejette comme les autres des pelotes muqueuses riches en algue mais
contrairement aux autres espèces observées la majorité des algues y sont endommagées
ou sous forme de débris et donc digestibles par l'Actinie. Dans cette espèce, une
substance provoquant l'altération des zooxanthelles a été mise en évidence dans les
tentacules, la collerette et le disque oral. Cette substance est active sur d'autres
espèces d'Anthozoaires : Aiptasia et Zoanthus, mais n'existe pas chez ces espèces.
Cette substance en provoquant l'augmentation du nombre des algues altérées permet à
Phyllactis flosculifera d'utiliser pour son alimentation la majorité des algues
expulsées des tissus.

LES ZOOXANTHELLES PRODUCTEURS PRIMAIRES

La première démonstration d'un transfert de substances élaborées par photosynthèse
de l'algue vers les tissus d'un Actiniaire a été fournie par Muscatine Hand (1958)
pour Anthopleura elegantissima. La technique d'autoradiographie employée n'a toutefois
pas permis d'identifier les produits qui étaient fournis à l'hôte. Trench (1971)
estime à 50 % le ^{14}C produit par photosynthèse et fourni par l'algue aux tissus
d'Anthopleura. Taylor (1968) estime ce transfert à 60 % pour Anemonia sulcata et
Von Holt et Von Holt (1968) à 25 % pour Condylactis gigantea. Ces produits élaborés
dans les xanthelles sont incorporés plus particulièrement dans les protéines et les
lipides de l'hôte.
Les zooxanthelles provenant des tissus d'Anthopleura récemment isolées par centrifugation et mises à incuber en présence de bicarbonate marqué voient leur photosynthèse augmentée en présence d'homogénat de tissu en provenance de l'hôte.
L'excrétion dans le milieu des produits marqués au ^{14}C est également stimulée
(Trench, 1971, c). Si l'homogénat est bouilli ou s'il provient d'un hôte préalablement
débarassé de ses symbiotes (aposymbiotique), l'effet est réduit de plus de 50 %. Il
semble donc exister un facteur de stimulation propre au tissu. Les produits transférés
de l'algue vers les tissus de l'hôte sont différents de ceux qui restent dans la
cellule ce qui permet de démontrer que les produits excrétés ne proviennent pas
d'une lyse des cellules in vitro. Le transfert est donc sélectif. Lewis et Smith (1971)
ont utilisé pour l'étude de ces phénomènes une technique d'inhibition du transfert
in situ des produits marqués élaborés par le symbiote. En présence de bicarbonate
marqué et en ajoutant le produit froid dont on souhaite mettre en évidence le transfert,
le produit transféré se retrouve dans le milieu. Cette technique a permis de mettre
en évidence la production par le symbiote de glycine, d'alanine et de glucose.
L'addition d'éthylène glycol, de sorbose, de fructose, de glucosamine et de divers
acides aminés n'induit pas l'apparition de ces mêmes produits dans le milieu.

CONCLUSION

Il était classiquement admis que les Actiniaires se nourrissaient du plancton et
de proies microscopiques. Dans le cas où ils hébergeaient des symbiotes il était
également admis que ces symbiotes intervenaient dans la nutrition. Il semble
maintenant que la part revenant aux matières organiques en suspension ou en solution
soit très importante au moins chez les Cérianthaires. La symbiose des Actiniaires
paraît efficace. Il existe même des procédés de régulation commandés par le symbiote
qui règlent la phototaxie des Anthopleura elegantissima (Pearse, 1974). Il existe
également des passages de métabolites de l'hôte vers l'algue (Cook, 1971). Des
travaux récents concernant la physiologie nutritionnelle des coraux sorientent
vers le métabolisme des lipides (Benson, Patton et Abraham, 1978). A notre connais-
sance cette voie vient juste d'être abordée chez les Actiniaires (Van Praët, 1978)
et pas dutout chez les Cérianthaires, mais elle mérite peut être d'être suivie et
approfondie.

REFERENCES

Afzelius, B. A., et B. Rosen (1965). Nutritive phagocytosis in animal cells. An
    electron microscopical study of the gatroderm of the hydroid Clava squamata.
    Z. Zellf. Mikr. Anat., 67, 24-33.
Benson, A. A., J.S. Patton et S. Abraham (1978). Energy exchange in coral reef
    ecosys tems. Atoll Research Bulletin. 220, 33-53.
Chapeaux, M. (1972). Recherches sur la digestion des Coelentérés. Archs. Zool.
    Expér. gén., 3, 279-283.
Chia, F.S. (1972). Note on the assimilation of glucose and glycine from sea water
    by the embryos of the sea anemone Actinia equina. Can. J. Zool., 50, 1333-1334.
Cook, C.R. (1971). Transfer of ^{35}S-labeled material from food ingested by Aiptasia
    to its endosymbiotic zooxanthellae. In H. M. Lenhoff, L. Muscatine, L. V. Davis
    (Eds.), Experimental Coelenterate Biology, University of Hawai Press, Honolulu,
    pp. 218-224.
Droop. M. R. (1963). Algea and invertebrates in symbiosis. Symp. Soc. gen. Microbiol.,
    13, 171-199.
Gibson, D. et G. H. Dixon (1969). Chymotrypsin like proteases from the sea anemone
    Metridium senile. Nature, 222, 753-756.
Jeuniaux, C. (1962). Digestion de la chitine chez les Actiniaires. Cah. Biol. Mar.,
    3, 391-400.
Jordan, H. (1907). Die Verdauung bei den aktinien. Pflüger Arch. fur. ges. Physiol.,
    116, 617-624.
Jorgensen, C. B. (1976). August Pütter, August Krogh and modern ideas on the use of
    dissolved organic matter in aquatic environment. Biol. Rev., 51, 291-32.
Krijgsam, B. J. et F. H. Talbot (1953). Experiments on digestion in sea anemone.
    Archs. Int. Physiol. Biochim., 61, 277-291.
Krukenberg, C. F. W. (1878). Über den Verdauungsmodus der Aktinnen. Vergl. Physiol.
    1, 33-56.
Lewis, D. H. et D. C. Smith (1971). The autotrophic nutrition of symbiotic marine
    Coelenterates with particuliar reference to the hermatypic corals. Proc. Roy. Soc.
    B. London. 178, 119-129.
Lindstedt, K. J., Muscatine, L. et H. M. Lenhoff (1968). Valine activation of
    feeding in the sea anemone Boloceroides. Comp. Biochem. Physiol. 26, 567-572.
Lindstedt, K. J. (1971). Biphasic feeding response in a sea anemone : control by
    asparagine and glutathione. Science, N. Y., 173, 333-334.
McFarlane, I. D. (1970). Control of preparatory feeding behaviour in the sea anemone
    Tealia. J. Exp. Biol., 53, 211-220.
McLaughlin, J. J. et P. A. Zahl (1966). Endozoic algae. In S. M. Henry (Ed) Symbiosis I
    Academic Press, New York. pp. 257-297.

Mesnil, F. (1901). Recherches sur la digestion intracellulaire et les diastases des Actinies. Annls. Inst. Pasteur, Paris, 15, 352-397.

Mouchet, S. (1929). Présence de xanthine chez les Actinies. Bull. Soc. Zool. France, 54, 345-350.

Murdock, G. R. (1971). The formation and assimilation of alcohol soluble proteins during intracellular digestion by Hydra littoralis and Aiptasia. In H. M. Lenhoff, L. Muscatine, L. V. Davis (Eds.), Experimental Coelenterate Biology, University of Hawai Press, Honolulu. pp. 137-145.

Muscatine, J. et C. Hand (1958). Direct evidence for the transfer of materials from the symbiotic algae to the tissued of a coelenterate. Proc. Nat. Acad. Sci. U.S.A., 444, 1259-1263.

Needler Arai, M. et G. L. Walder (1973). The feeding response of Pachycerianthus fimbriatus. Comp. Biochem. Physiol., 44A, 1085-1092.

Nicol, J. A. C. (1959). Digestion in sea anemones. J. Mar. Biol. Assoc. U. K., 38, 469-476.

Pantin, C. F. A. (1950). Behaviour pattern in lower invertebrates. Symp. Soc. Exp. Biol., 4, 175-195.

Pearse, V. B. (1974). Modification of seaanemone behaviour by symbiotic zooxanthellae Phototaxis. Bull. mar. Biol. Assoc. Woods Hole, 147, 630-640.

Schlichter, D. (1974). Aufnahme in Neerwasse gelöster Aminosäuren durch Anemonia sulcata. Dans une terschiedliche resorptions vermögen von Ekto- und Endoserm. Z. Morphol. Okol. Tiere Dtsch., 79 (1), 385-405.

Schlichter, D. (1975). The importance of dissolved organic compounds in sea water for the nutrition of Anemonia sulcata. In H. Barnes (E. D.), The biochemistry, physiology and behaviour or marine organisms in relation to their ecology. Aberdeen University Press, pp. 395-405.

Schlichter, D. (1978). On the ability of Anemonia sulcata to absorb charged and neutral amino, acids simultaneously. Mar. Biol. 45, 97-104.

Steele, R. D. (1976). Light intensity as a factor in the regulation of the density of sybiotic zooxanthellae in Aiptasia tagetes. J. Zool. London, 179, 387-405.

Steele, R. D. et N. I. Goreau (1977). The breakdown of symbiotic zooxanthellea in the sea anemone Phyllactis flosculifera. J. Zool. London, 181, 421-437.

Taylor, D. L. (1968). In situ studies on the cytochemistry and ultrastructure of a symbiotic marin dinoflagellate. J. mar. Biol. Assoc. U. K., 48, 349-366.

Taylor, D. L. (1973). Cellular interactions of algal invertebrates symbiosis. Adv. mar. Biol. 11, 1-56.

Tiffon, Y. (1973). Latency and sedimentability of acid hydrolases in sterile sept homogenates of Cerianthus lloydi. Comp. Biochem. Physiol., 45 B, 731-740.

Tiffon, Y. et M. Daireaux (1974). Phagocytose et pinocytose par l'ectoderme et l'endoderme de Cerianthus lloydi. J. exp. mar. Biol. Ecol., 16, 155-165.

Tiffon, Y. et J. Bouillon (1975). Digestion extracellulaire dans la cavité gastrique de Cerianthus lloydi. Structure du gastroderme, localisation et propriétés des enzymes protéolytiques. J. exp. mar. Biol. Ecol., 18, 255-269.

Tiffon, Y. et J. S. Hugon (1977). Localisation ultrastructurale de la phosphatase acide et de la phosphatase alcaline dans les cloisons septale stériles de L'Anthozoaire Pachycerianthus finbriatus. Histochem., 54, 289-297.

Trench, R. K. (1971 a). The physiology and biochemistry of zooxanthellea symbiotic with marine coelenterates. I. Assimilatoon of photosynthetic products of zooxanthellae by two marine coelenterates. Proc. Roy. Soc. London, B. 177, 225-235.

Trench, R. K. (1971 b). The physiology and biochemistry of zooxanthellae symbiotic with marine coelenterates. II. Liberation of fixed ^{14}C by zooxanthellae in vitro. Pro. Roy. Soc. London, B 177, 237-250.

Trench, R. K. (1971 c). The physiology and biochemistry of zooxanthellae symbiotic with marine coelenterates. III. The effect of homogenates of host tissues on the excretion of photosynthetic products in vitro by zooxanthellae from two marine coelenterates. Proc. Roy. Soc. London, B. 177, 251-254.

Vader, W. et S. Lönning (1975). The ultrastructure of the mesenterial filaments of
   Bolocera tuediae. Sarsia 58, 79-88.
Van Praët, M. (1976). Les activités phosphatasiques acides chez Actinia equina et
   Cereus pedunculatus. Bull. Soc. Zool. 101, 367-376.
Van Praët, M. (1977). Les cellules à concrétions d'Actinia equina. C. R. Ac. Sc. Paris
   285, 45-48.
Van Praët, M. (1978). Etude histochimique et ultrastructurale des zones digestives
   d'Actinia equina. Cah. Biol. Mar. 19, 415-432.
Von Holt, C. et N. Von Holt (1968). Transfer of photosynthetic products from zooxan-
   thellae to Coelenterates hosts. Comp. Biochem. Physiol., 24, 73-81.
Williams, R. B. (1972). Chemical control of feeding behaviour in the sea anemone
   Diadumene luciae. Comp. Biochem. Physiol., 41A, 361-371.

THE PHOSPHORUS ECONOMY OF
THE GREEN HYDRA SYMBIOSIS

F. P. WILKERSON

Department of Botany, The University, Bristol, U.K.

ABSTRACT

More phosphate was found in symbiotic green hydra than aposymbiotic. Within hydra, there was more phosphate in the algal component than the animal. Starved green hydra contained far less phosphate than those fed on Artemia nauplii. The phosphate content of the nauplii was more than sufficient to meet the requirements of the hydra. Green hydra absorbed more radioactively labelled orthophosphate from the medium than aposymbionts and a greater percentage of the ^{32}P taken up was found in the algal component. Light did not affect phosphate uptake. Addition of phosphate to the growth medium did not increase the budding rate or total number of algae in either fed or starved animals. The presence of the symbiotic algae, their participation in possible recycling of phosphate within the symbiosis and the subsequent advantages to animals living in habitats low or devoid of phosphate is discussed.

KEY WORDS

Symbiosis; green hydra; inorganic nutrients; phosphate; recycling; productivity.

INTRODUCTION

Very many problems in symbiosis involve nutrition either directly or indirectly (Droop, 1963). Extensive studies on associations between invertebrates and endo-symbiotic algae have demonstrated that algae release soluble photosynthate to their hosts (Smith and others, 1969). However, Lewis and Smith (1971) have drawn attention to the possibility that other nutrients could also move between algae and host. This could be particularly important to organisms such as reefbuilding corals living in tropical waters poor in both suspended particulate and dissolved nutrients. Some essential nutrients such as nitrogen and phosphorus may be obtained by feeding on zooplankton or by direct uptake of nutrients dissolved in the surrounding water; corals obtain some but not all the phosphorus they require by net uptake of dissolved inorganic phosphorus at low environmental concentrations (D'Elia, 1977). However, the high incidence of symbiosis in habitats low in available nitrogen and phosphorus raises the suggestion of recycling mechanisms between host and symbionts since there may be conservation of nutrients by minimising excretory losses. Geddes (1882) was the first to propose that symbiotic algae performed an intracellular 'renal' function. Algae can store much more phosphorus than animals within their vacuoles or as inert polyphosphate granules. This store could be released in times of starvation to the animal.

It is difficult to make a vigorous and direct demonstration that substances move from animal to symbiont. Yonge and Nicholls (1931) proposed that zooxanthellae utilise inorganic phosphate produced by host tissue although direct proof of this is still lacking. The studies of Pütter (1911), Kawagata (1953), and Cates and McLaughlin (1979) suggest that zooxanthellae remove ammonia produced by the host. Cook (1971) showed that the zooxanthellae of Aiptaisia could obtain labelled sulphate from the actinians food.

So a certain amount of evidence is available to show that endosymbiotic algae can obtain nutrients from the animal in the form of waste materials from the host and transference of ingested food. Although not experimentally proved, it is reasonable to expect that in lower animals with symbionts, nutrients dissolved in the surrounding water might find their way to the symbiont.

The purpose of this investigation was to evaluate the relationship between phosphate nutrition and endosymbiotic algae and to establish its importance in symbiosis. Actual levels of phosphate in the environment and animals were measured and the flux of dissolved inorganic radioactive phosphate into the animals investigated. The effect of added phosphate to the culture medium on growth was observed.

MATERIALS AND METHODS

Experimental organisms

Specimens of green hydra 'European' strain were donated by Dr. L. Muscatine. This strain has been in culture over 12 years. The 'Jubilee' strain was obtained from Professor D.C. Smith and was originally isolated in 1977 from the River Frome where the river enters the eastern end of Oldbury Court Estate, Bristol, U.K. Aposymbiotic forms of each strain were produced by pretreating the green hydra in 10^{-5}M DCMU for 1 week and then exposing them to bright light as described by Pardy (1976). These were cultured in 'M' solution (Muscatine & Lenhoff, 1965) but minus tris buffer at 15°C, 1200 lux on a 12h light/dark cycle. The animals were fed 3 times a week on freshly hatched Artemia nauplii.

Separation of algae from animal tissues

The hydra were homogenised in a glass tissue grinder and the homogenate centrifuged at 3000 rpm for 4 mins. The supernatant was collected, the pellet washed and centrifuged again. The pellet contained all the algal cells and the supernatant and washings were combined to form the animal fraction.

Assay of phosphate

Hydra and Artemia nauplii. 100 hydra or 1000 Artemia nauplii were homogenised with a tissue grinder. The homogenate was digested using 5% potassium persulphate and then autoclaved at 15lbs/sq.in. for 30 min (Menzel & Corwin, 1965). Some homogenates were separated into algal and animal components as described above, and also digested. The total phosphate was measured spectrophotometrically using the method of Murphy and Riley (1962) modified by the addition of 95% ethanol just before adding the phosphate reagent.

River water. 60 ml samples were collected in plastic bottles and filtered through a 0.45 µm Millipore filter. The total phosphate present was then measured using the modified Murphy and Riley (1962) method as above.

Measurement of radioactive phosphate uptake

Twenty hydra were placed in a glass vial containing 5ml of hydra medium and 1-2µCi of carrier-free ^{32}P labelled orthophosphate. After 5 hr at $15^{o}C$ in the light, the hydra were thoroughly washed in hydra medium and the radioactivity measured in whole animals, algal and animal fractions, using Cerenkov counting in a Nuclear Chicago Isocap 300 scintillation counter. Radioactivity in the remaining medium was also estimated.

Protein estimation

The protein content of 2ml hydra homogenates, supernatants or algal pellets was assayed by the procedure of Lowry and others (1951). Results are expressed per hydra or per microgram of protein, the latter allowing for different sizes of animal in the sample.

Determination of growth by budding rate

10 hydra each with 1 developed bud were put in 20ml of hydra medium or medium supplemented with 1mM KH_2PO_4. The hydra were fed 3 times a week on Artemia nauplii and the number of feeding hydranths in each dish were counted daily. Some hydra were starved for comparison. A semi-log plot of the number of feeding hydranths against time was plotted and a straight line drawn so as to best fit the data (Fig.1). Then the standard logarithmic growth rate equation was applied and the rate of log growth (K) calculated according to Loomis (1954).

Total algal population

10 hydra were homogenised and the total number of algae estimated using a Neubauer improved haemocytometer.

RESULTS

Phosphate assays

The results of the chemical assays are shown in Table 1.

TABLE 1. The chemical estimation of total phosphate

	µg/L total phosphate
River Frome	46 ± 3

	ng/hydra
Fed green hydra ('European' strain)	47 ± 4
Fed aposymbionts ('European' strain)	30 ± 4
Pellet (algal)	21 ± 3
Supernatant (animal)	14 ± 0.5
% in pellet = 67%	
4 day starved green	19 ± 0.4

	ng/nauplius
Freshly hatched Artemia	24 ± 2

Mean ± S.E.M.

The level of total soluble phosphate in the natural environment of the hydra is only 50µg/L. There was over 50% more phosphate in green hydra than in aposymbionts. In green hydra, 60-70% of the phosphate is in the pellet, which is predominantly algal in composition. If the green hydra are starved the amount of phosphate they contain declines. After 4 days starvation with no Artemia nauplii this reduction reached 50%. The levels of phosphate in Artemia are relatively large, and must therefore be quantitatively a very important source for fed hydra. It is not known how much of the Artemia-derived phosphate actually enters or stays in the symbiotic association.

Phosphate uptake from media

Hydra readily take up ^{32}P labelled orthophosphate from the medium. Green hydra took up more than twice as much radioactive phosphate per unit protein than the aposymbiotic (Table 2).

TABLE 2. Phosphate uptake by hydra after 5hrs incubation

	Net dpm/μg protein
European Green	510 $\pm$ 64
European Aposymbiotic	217 $\pm$ 17
Green:Albino = 2.4:1	
Jubilee pellet (algal)	82 $\pm$ 6
Jubilee supernatant (animal)	36 $\pm$ 5
% in pellet = 70%	
Jubilee green in light	81 $\pm$ 6
Jubilee green in dark	76 $\pm$ 5

Mean $\pm$ S.E.M.

In green hydra the pellet (algal fraction) contained 70% of the radioactive phosphate (Table 2). The effect of darkness on phosphate uptake was investigated, since this reduces phosphate uptake in many plant systems, but there was no significant effect of darkness on uptake from the media.

The effect of phosphate on budding rate.

The hydra medium in which laboratory cultures of hydra are maintained contains no phosphate. The results of adding 1mM phosphate to the medium of green hydra is shown in Fig. 1.

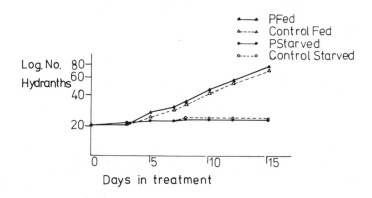

Fig. 1. The effect of 1mM PO_4 on budding by green hydra.

There is no significant effect of phosphate on budding rate either in fed or starved hydra. There was also no significant effect of adding 1mM phosphate to the albino cultures (Table 3).

TABLE 3. The effect of 1mM PO_4 on budding rate by albino hydra

| | | K (growth constant) | |
		Control	Phosphate
European	Fed	0.1	0.1
	Starved	0.04	0.03
Jubilee	Fed	0.09	0.07
	Starved	0.01	0.01

The effect of phosphate on total algal population

Addition of phosphate was shown by Pool (1976) to increase the total algal population in starved green hydra of the Florida strain. No such effect was found here with the Jubilee strain of green hydra (Table 4).

TABLE 4. The effect of 1mM PO_4 on total algal number in green hydra (Jubilee)

| | Total algae $x10^4$/µg animal protein | |
	Control	Phosphate
Fed	7.1 ± 0.3	6.8 ± 0.7
Starved	6.4 ± 0.7	7.0 ± 0.8

Mean ± S.E.M.

DISCUSSION

Hydra can take up dissolved phosphate. Since green hydra both take up and contain more phosphate than aposymbiotic hydra it is likely that endosymbiotic algae play an important role in both the storage of phosphate and its acquisition from the environment.

The habitat in which green hydra is found has a level of 50 µg/L total soluble phosphate. Such low levels may be important in limiting the total productivity of the stream ecosystem. The phosphate content of the hydra per unit volume is almost 300-400 times higher. However, the relatively high phosphate content of Artemia nauplii suggests that the stream zooplankton are a much richer source of phosphate than the surrounding water. The phosphate content of one Artemia nauplius is half that of an entire hydra and most hydra eat 6-7 nauplii readily at each feeding.

When food is present in the habitat excess phosphate derived from feeding on zoo-plankton could be moved to the algae and stored. During subsequent starvation periods, such stored phosphate could then pass back from the algae to the host. Although phosphate could be absorbed from solution, there was no evidence that adding phosphate to hydra medium (normally phosphate free) increases the rate of growth, even in starved hydra. There could be two reasons for this, not mutually exclusive. Firstly, the residual of stored phosphate even in starved hydra could still have been sufficient to maintain growth. Secondly, it could be not just phosphate, but a balance of nutrients which is required by starved hydra. Although phosphate may frequently limit the productivity of an ecosystem, it is likely that this effect only concerns the primary producers at the base of the food chain. The overall amount of food passing along the chain is reduced but key nutrients such as nitrogen and phosphorus move up in a balanced way. For organisms higher up the chain all forms of nutrient are in short supply (including those not originally limiting production). Thus, the role of symbiotic algae in storing and recycling nutrients will be important if they are recycling either a balanced range of nutrients or if some nutrients which would otherwise have been lost as waste products (e.g. nitrogen compounds) are converted back to useful forms.

The high incidence of symbiotic associations in low nutrient waters suggests that such recycling is nevertheless important.

ACKNOWLEDGEMENTS

This work was financed by a Natural Environmental Research Council Studentship.

REFERENCES

Cates, N., and J.J.A. McLaughlin (1979). Nutrient availability for zooxanthellae derived from physiological activities of Condylactus spp. J. exp. mar. Biol. Ecol., 37, 31-41.

Cook, C.B. (1976). Sulphate utilisation in green hydra. In G.O. Mackie (Ed.), Coelenterate Ecology and Behaviour, Plenum Press, New York. pp.415-422.

D'Elia, C.F. (1977). The uptake and release of dissolved phosphorus by reef corals. Limnol. Oceanogr., 22, 301-315.

Droop, M.R. (1963). Algae and invertebrates in symbiosis. Symp. Soc. Gen. Microbiol., 13, 171-199.

Geddes, P. (1882). On the nature and function of the 'yellow cells' of radio-larians and coelenterates. Proc. Roy. Soc. Edinb. B, 11, 377.

Kawaguti, S. (1953). Ammonium metabolism of the reef corals. Biol. Okayama Univ., 1, 171-176.

Lewis, D.H., and D.C. Smith (1971). The autotrophic nutrition of symbiotic marine coelenterates with special reference to hermatypic corals. I. Movement of photosynthetic products between the symbionts. Proc. Roy. Soc. London B, 178, 111-129.

Loomis, W.J. (1954). Environmental factors controlling growth in hydra. J. Exp. Zool., 126, 223-234.

Lowry, O., N. Rosebrough, A. Farr and R. Randall (1951). Protein measurement with the Folin phenol reagent. J. Biol. Chem., 193, 265-275.

Menzel, D.W. and N. Corwin (1965). The measurement of total phosphorus in sea water based on the liberation of organically bound fractions by persulphate oxidation. Limnol. Oceanogr., 10, 280-282.

Murphy, J. and J.P. Riley (1962). A modified single solution method for the determination of phosphate in natural waters. Anal. Chim. Acta., 27, 31-36.

Muscatine, L., and H.M. Lenhoff (1965). Symbiosis of hydra and algae. II. Effects of some environmental cations on growth of symbiotic and aposymbiotic hydra. Biol. Bull. mar. biol. Lab. Woods Hole, 128, 415-424.

Pardy, R.L. (1976). The production of aposymbiotic hydra by the photodestruction of green hydra zoochlorellae. Biol. Bull. Mar. biol. Lab. Woods Hole, 151, 225-235.

Pool, R.R. (1976). Symbiosis of Chlorella and Chlorohydra viridissima. Ph.D. dissertation, University of California, Los Angeles.

Pütter, A. (1911). Der Stoffwechsel der Aktinien. Z. allg. Physiol., 12, 297-322.

Smith, D.C., L. Muscatine and D.H. Lewis (1969). Carbohydrate movement from autotrophs to heterotrophs in parasitic and mutualistic symbiosis. Biol. Rev., 44, 17-90.

Yonge, C.M., and A.G. Nicholls (1931). Studies on the physiology of corals. IV. The structure, distribution and physiology of zooxanthellae. Sci. Rept. Gt. Barrier Reef Exped., 1, 135-176.

NUTRITION OF MARINE SPONGES. INVOLVEMENT OF SYMBIOTIC BACTERIA IN THE UPTAKE OF DISSOLVED CARBON

C. WILKINSON* AND R. GARRONE**

*School of Environ. Life Sc., Murdoch Univ., Murdoch,
Western Australia
**Lab. Histology, Univ. Claude Bernard, 69622 Villeurbanne, France

ABSTRACT

The uptake of tritium labelled proline from the ambient water by the marine sponge Chondrosia reniformis was studied by electron microscope radioautography. The dissolved amino acid was very fastly incorporated into symbiotic bacteria in the sponge, and more slowly into sponge cells. The ability to use dissolved organic carbon may be important to sponges in water with low levels of particulate carbon.

KEYWORDS

Sponge nutrition; symbiotic bacteria; nutrient translocation; radioautography.

INTRODUCTION

It is well recognized that marine sponges efficiently filter particulate matter from the ambient water (Jørgensen, 1966 and 1976; Reiswig, 1971). The hypothesis that sponges may also be able to utilize dissolved organic carbon was advanced by Pütter in 1914. Proof that sponges can remove some dissolved compounds was given by Stephens and Schinske (1961) who showed in aquarium studies that two sponge species removed the amino acid glycine when present in a large concentration. Reiswig (1971), in a detailed study of tropical sponge nutrition in situ, showed that in one species of Verongia, the retention of particulate matter was insufficient for sponge metabolism. He inferred that utilization of dissolved compounds accounted for this discrepancy and suggested that the large population of bacteria within this sponge may be the mechanism for uptake of dissolved carbon.

Neither the results of aquarium studies by Stephens and Schinske (1961) using high concentrations of glycine nor the speculation advanced by Reiswig (1971) have been verified or tested. In this preliminary paper, the fate of the amino acid proline fed to the sponge Chondrosia reniformis is reported. The tritium labelled proline was administered in an attempt to trace the synthesis of collagen, which is prominent in this sponge (Garrone, 1978).

MATERIALS AND METHODS

Chondrosia reniformis Nardo specimens were collected from the mediterranean sea
(Marseille, Sète and Banyuls) and maintained in circulating sea water until used.
For incorporation experiments, small specimens (1 to 2 cm^3) were placed in 50 cm^3
of natural sea water. Tritium labelled proline (specific activity: 23 Ci/mM, CEA,
France) was added to give a final concentration of 50 µCi/cm^3. In some cases,
cycloheximide was added (at a concentration of 150 µg/cm^3) to inhibit cell protein
synthesis. Specimens were removed from the reaction mixture after 15 mn to 4 h,
and small sections were fixed for electron microscopy in 2 % glutaraldehyde in
0.1M sodium cacodylate containing unlabelled proline, post-fixed in osmium
tetroxide, dehydrated in ethanol and embedded in Epon resin. Thin sections were
stained with uranyl acetate and lead citrate. For radioautography, the sections,
coated with an "Ilford L4" nuclear emulsion, were stored for 4 to 6 weeks and
developed either with a gold latensification-elon ascorbic acid developer or with
a phenidon developer. Observations were carried out on "Hitachi HU 12A" and
"Philips EM 300" electron microscopes.

RESULTS AND DISCUSSION

Like many marine sponges (Lévi and Lévi, 1965; Vacelet, 1975; Wilkinson, 1978a),
Chondrosia reniformis contains a large population of symbiotic bacteria (Fig. 1).
When individuals of this sponge were incubated in sea water containing tritium
labelled proline, the amino acid was removed from the water and incorporated into
the sponge. Short incubation times (15 mn to 1 h) allowed the detection of
radioactivity within the bacteria alone. Longer incubation times were necessary to
significantly label the cells and the intercellular matrix (see Garrone, 1978).
At any time of incubation, however, the uptake of proline by bacteria was
demonstrated by radioautography (Fig. 2). Inhibition of cell protein synthesis by
cycloheximide led to an increased radioactivity of the bacteria (Fig. 3) indicating
that the label of the bacteria cannot be interpreted as the result of cell
activity (for example, the release of a labelled protein which was then digested
by the bacteria) and must be considered as the result of a direct uptake. In other
respects, it could be clearly seen that some labelled bacteria were ingested by
sponge cells (Fig. 4).

Previous studies on collagen biosynthesis in sponges (Garrone, 1978) and these
results demonstrate that sponges can absorb dissolved proline. The amino acid
being incorporated in sponge cells and sponge symbiotic bacteria, it can be thus
concluded that both sponge cells and symbiotic bacteria possess uptake systems for
this amino acid.

Previously, Stephens and Schinske (1961) showed incorporation of glycine into two
sponge species, Microciona prolifera and Cliona celata. However, they did not
differentiate whether the amino acid was taken up by the sponge or by symbionts
within the sponge. Madri and others (1971) examined bacterial populations from
Microciona prolifera, but as the bacterial populations reported in the sponge and
ambient water were unrealistically low, the substantial part of their reported
sponge populations were considered unreliable (Wilkinson, 1978b). Reiswig (1971)
suggested that the large tropical sponge Verongia gigantea was able to make up the
deficiency between the required organic carbon compared to the available
particulate carbon by incorporating dissolved organic carbon. In view of the
results reported here, his suggestion that symbiotic bacteria were able to
incorporate dissolved organic carbon for the sponge is credible.

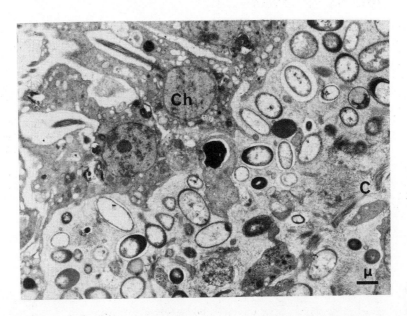

Fig. 1. Numerous bacteria are contained in the intercellular matrix of Chondrosia, especially in the choanosome area. Ch: choanocytes; C: collagen fibrils.

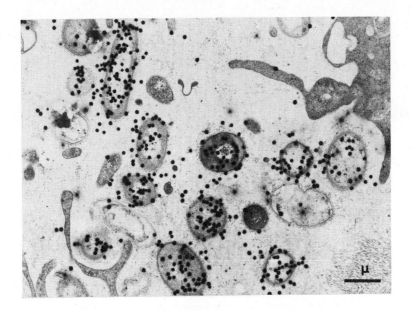

Fig. 2. Labelled bacteria after ³H-proline incorporation. The label appeared as a halo of silver grains on and around the bacteria (phenidon developer).

C. Wilkinson and R. Garrone

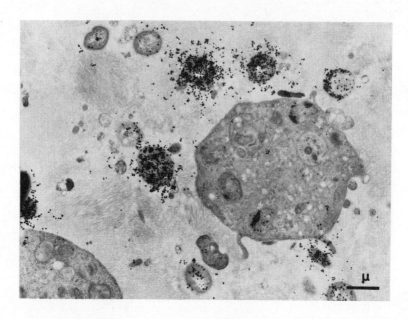

Fig. 3. Heavily labelled bacteria after ³H-proline incorporation and cycloheximide incubation (gold latensification-elon ascorbic acid developer).

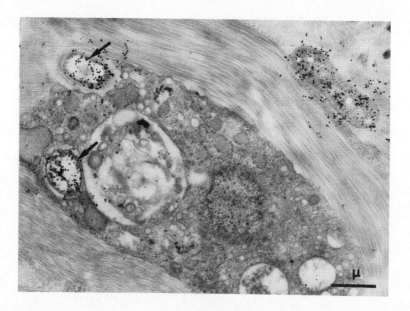

Fig. 4. Ingestion of radioactive bacteria (arrows) by a sponge cell (gold latensification-elon ascorbic acid developer).

However, little is known of the role of symbiotic bacteria in sponges. Chondrosia reniformis has been shown to contain similar sponge-specific symbionts to those isolated from Great Barrier Reef sponges (Wilkinson, 1978b and unpublished results) which were believed to be implicated in nitrogen recycling within the host sponge.

These preliminary findings demonstrate that sponge bacteria can incorporate the amino acid proline, and further experiments are required to determine the nutritional significance of dissolved organic carbon uptake in marine sponges. The transfer of significant amounts of dissolved organic carbon to the host sponge from the symbiotic bacteria would allow some sponges to survive in areas with low levels of particulate organic carbon.

ACKNOWLEDGEMENTS

The authors would like to express their gratitude to the "Centre de Microscopie Electronique Appliquée à la Biologie et à la Géologie" of the University Claude Bernard for technical facilities. This work was supported in part by the C.N.R.S. (L.A. 244).

REFERENCES

Garrone, R. (1978). Phylogenesis of connective tissue. Morphological aspects and biosynthesis of sponge intercellular matrix. In L. Robert (Ed.), Front. Matrix Biol., Vol. 5. S. Karger, Basel.

Jørgensen, C. B. (1966). Biology of suspension feeding. Pergamon Press, London.

Jørgensen, C. B. (1976). August Pütter, August Krogh, and modern ideas on the use of dissolved organic matter in aquatic environments. Biol. Rev., 51, 291-328.

Lévi, C., and Lévi, P. (1965). Populations bactériennes dans les Eponges. J. Microscopie, 4, 151.

Madri, P. P., Hermel, M., and Claus, G. (1971). The microbial flora of the sponge Microciona prolifera Verrill and its ecological implications. Botanica Mar., 14, 1-5.

Pütter, A. F. (1914). Der Stoffwechsel der Kieselschwämme. Z. Allg. Physiol., 16, 65-114.

Reiswig, H. M. (1971). Particle feeding in natural populations of three marine demosponges. Biol. Bull., 141, 568-591.

Stephens, G. C., and Schinske, R. A. (1961). Uptake of amino acids by marine invertebrates. Limnol. Oceanogr., 6, 175-181.

Vacelet, J. (1975). Etude en microscopie électronique de l'association entre bactéries et spongiaires du genre Verongia (Dictyoceratida). J. Micr. Biol. Cell., 23, 271-288.

Wilkinson, C. R. (1978a). Microbial associations in sponges. III. Ultrastructure of the in situ associations in coral reef sponges. Mar. Biol., 49, 177-185.

Wilkinson, C. R. (1978b). Microbial associations in sponges. II. Numerical analysis of sponge and water bacterial populations. Mar. Biol., 49, 169-176.

KINETIC AND MORPHOLOGICAL ASPECTS OF PARTICLE INGESTION BY THE FRESHWATER SPONGE *EPHYDATIA FLUVIATILIS* L.

Ph. WILLENZ

Laboratoire de Biologie animale et Cellulaire, Université libre
de Bruxelles, 50, Av. F. Roosevelt, 1050 Bruxelles, Belgium

ABSTRACT

The transit of particles during nutrition of the freshwater sponge *Ephydatia fluviatilis* has been studied by feeding sponges with calibrated polystyrene latex beads (Diam. 0,794 µm). Beads have been added to the culture medium. After incubation times spreading from 5 minutes to 12 hours, sponges have been fixed for further ultrastructural studies. The results show that not only choanocytes and archaeocytes, but also pinacocytes manifest a phagocytic activity.

KEYWORDS

Sponge; latex beads; phagocytosis.

INTRODUCTION

Sponges are feeding by filtration of the inhaled water current produced by the flagellar activity of the choanocytes. These same cells take up nutritive particles by phagocytosis. Food is then transferred by a direct exchange to archaeocytes which digest and later provide nutritive substances to other specialized cells throughout the sponge.
Descriptions of the overall feeding systems of demosponges have been provided since the beginning of this century (van Tright, 1919; Pourbaix, 1931, 1932, 1933a, 1933b; van Weel, 1949; Kilian, 1952; Schmidt, 1970; Reiswig, 1971a, 1971b). Some of them were recently outlined by Frost (1976) in a review of sponge feeding.

The feeding role of pinacocytes is controversed. Those cells make up the dermal membrane and line the canals of the aquiferous system. Most investigators (van Weel, 1949; Kilian, 1952; Pavans de Ceccatty, 1958; Simpson, 1963; Bagby, 1970; Reiswig, 1971b; Harrison, 1972a, 1972b.) agree upon the phagocytic ability of pinacocytes. Meanwhile, Schmidt using ultra-violet fluorescent dye, never observed phagocytosis of bacteria by the pinacocytes of *Ephydatia fluviatilis* cultivated in vitro. Phagocytosis of baker's yeast occured exceptionaly, but Schmidt considered it as a cleaning function of sponge's surface and not as a nutritional function.

The aim of this work is to study from a morphological point of view the transit of particles ingested by *Ephydatia fluviatilis*, and to estimate the pinacocyte phagocytic activity. As "food", we have used calibrated latex beads, for three reasons.

1. Filtration rates of these beads in *Ephydatia fluviatilis* is known from a previous work. (Willenz et Rasmont, 1979).
2. Latex beads are easy to recognize in electron microscopy.
3. Polystyrene is expected to remain unaffected by lysosomial digestion.

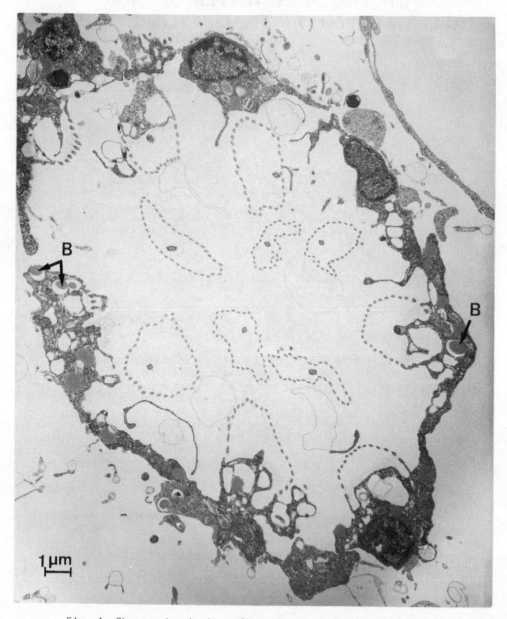

Fig. 1. Choanocyte-chamber after 10 min. incubation. B = bead.

MATERIAL AND METHODS

Ephydatia fluviatilis strain α (Van de Vyver 1970) raised from one gemmule,
cultivated at 20°C in Petri dishes containing mineral medium (Rasmont, 1961) have
been used. Ten days old sponges were incubated with latex beads 0,794 μm in
diameter at a concentration of 4.10^7 beads/ml of culture medium. (Dow Latex.
Serva Feinbiochemica.). During incubation, Petri dishes were stirred to maintain
an homogeneous repartition of these beads (45 cycles/min and 4 cm amplitude).

Sponges were fixed at room temperature after 9 different incubation times,
running from 5 minutes to 12 hours (Table 1). Fixation occured 1°) in 0,35%
glutaraldehyde in 0,025 M cacodylate buffer at pH 7,4 for 2 hours, 2°) in 1%
osmium tetroxyde in 0,025 M cacodylate buffer for 2 hours, after De Vos (1977).
After dehydratation in graded ethanol, sponges were embedded in ERL 4206 according
to Spurr (1969). Sections were obtained with a diamond knife on a Reichert OMU3
ultramicrotome. Before sectioning, the glass coverslips and the siliceous spicules
in the specimens were dissolved with a bath of concentrated hydrofluoric acid.
Thin sections were double stained with uranyl acetate and lead citrate (Reynolds,
1963) and examined in a AEI EM6B electron microscope at 60 Kv.

RESULTS

Beads Capture

Beads phagocytosis begins rapidly. Indeed after 10 minutes incubation, some of
them can already been observed in choanocyte-chambers (Fig. 1).

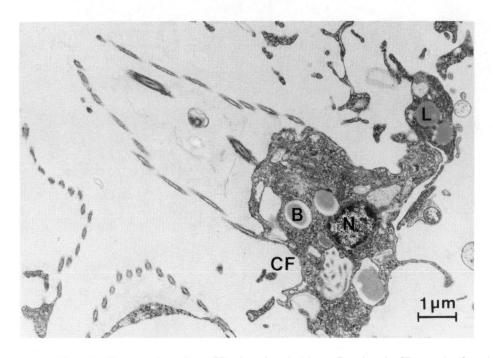

Fig. 2. Choanocyte after 20 min. incubation. B = bead, CF = cytoplasmic
fold, L = lipid droplets, N = nucleus.

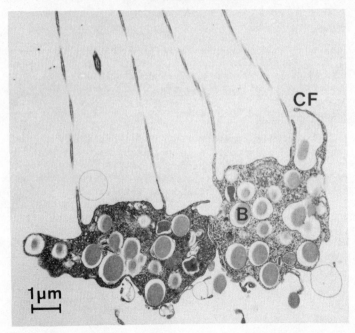

Fig. 3. Choanocyte after 40 min. incubation.
CF = cytoplasmic folds. B = bead.

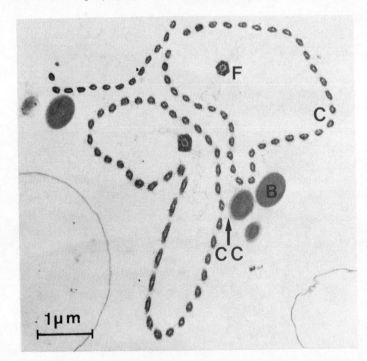

Fig. 4. Latex beads trapped in the collar cell coat.
B = bead. F = flagella. C = collar. CC = cell coat.

Figure 2 shows an isolated choanocyte in axial cut with three beads surrounded by their phagocytic membrane. The amount of phagosomes is increasing in course of time. After 40 minutes, large vacuoles resulting from the fusion of several phagosomes appear (Fig. 3). Note in this figure a bead undergoing phagocytosis, at the apical cytoplasmic fold.

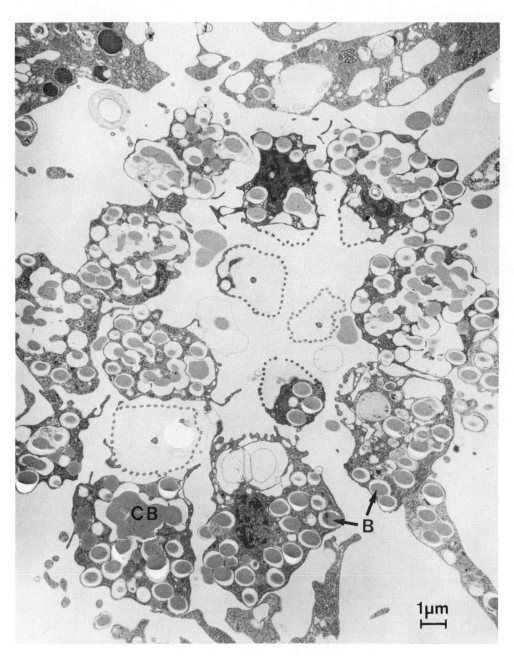

Fig. 5. Choanocyte-chamber after 5 hours incubation.
CB = coalescing beads. B = bead.

The examination of choanocyte collars leads to some interesting observations.
It seems that beads are very first trapped by the cell coat produced by the
cytoplasmic microvillies and the apex of the cell. Figure 4 (transversal cut in
a choanocyte collar) shows it very clearly.

Beads phagocytosis lasts for several hours. After 5 hours, all choanocytes are
highly filled with phagosomes which occupy almost all the cell volume. A lot of
these phagosomes have fused together building large vacuoles containing
coalescing beads (Fig. 5). Nevertheless this bead coalescence could be nothing
else than an artefact resulting from a partial dissolution of the polystyrene in
the embedding medium. Experiments to check this hypothesis are in progress.

Bead Transfer from Choanocytes to Archaeocytes

Archaeocytes are large wandering cells, characterized by a nucleolated nucleus.
Their great mobility brings them from time to time in close contact with
choanocyte-chambers. These contacts are very useful since they allow the
phagosome transit from choanocytes into archaeocytes, that is the digesting cells.

Archaeocytes observed below 1 hour incubation, even in close contact with
choanocytes never contain beads (Fig. 6.).

A sequence of pictures (fig. 7 to 11) all taken after 1 hour incubation allows to
demonstrate the process involved in the particle transit from choanocytes to
archaeocytes.

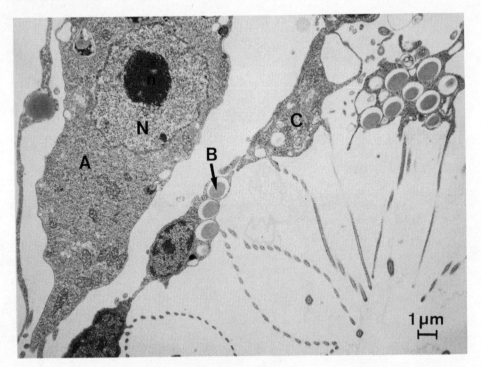

Fig. 6.   Cells below 1 hour incubation.
          A = archaeocyte. C = choanocyte. N = nucleus. n = nucleolus
          B = bead.

The first step consists in a very tight contact between both cell types followed
by a local fusion of the cell membranes at the level of a bead (fig. 7).
Secondly (Fig. 8 and 9), beads are intruded into archaeocytes. During that time
both cell types are deeply inserted in one another, sometimes leading to the
transfer of several beads together (Fig. 10).
Beads are then exocytosed by choanocyte and simultaneously phagocytosed by closely
joined archaeocytes (Fig. 11).

Finally, after 2 hours incubation, archaeocytes contain several beads, each of
them being surrounded by an individual membrane (Fig. 12).
As for choanocytes, as time goes on, the observed archaeocytic phagosomes fuse
together (Fig. 13).

Bead Release into the Mesohyl.

After 12 hours, beads are still present in the choanocytes. At that time some of
them are released directly into the mesohyl without being transferred into
archaeocytes (Fig. 14 and Fig. 15).

Beads and Pinacocytes

Pinacocytes make the dermal membrane and line the aquiferous canals. The dermal
membrane consists in a double layer of pinacocytes; the external is called
exopinacoderm, the internal, endopinacoderm. Both layers are separated by
collagen fibers.

When sponges are incubated in presence of latex beads, beads appear in the
exopinacoderm after 20 minutes and, in the collagen layer and the endopinacoderm
after 40 minutes (Fig. 16).
This timing gives an indication on the way of transit of beads which are obviously
ingested from the external medium by the exopinacocytes (Fig. 16) and later
transferred to the endopinacocytes (Fig. 17).

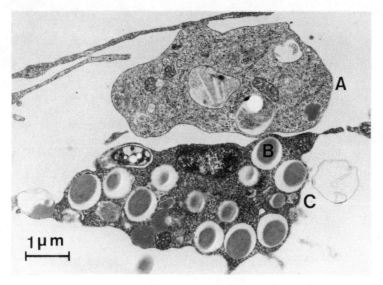

Fig. 7. Choanocyte and archaeocyte after 1 hour incubation.
        A = archaeocyte. C = choanocyte.
        B = bead.

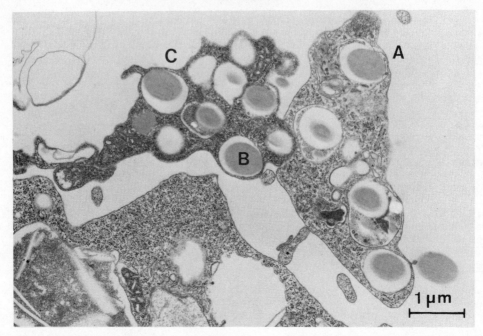

Fig. 8. Choanocyte and archaeocyte after 2 hours incubation.
        A = archaeocyte. C = choanocyte.
        B = bead.

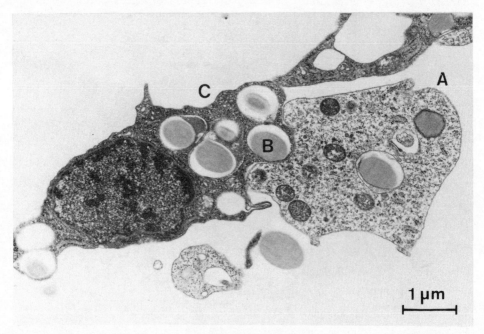

Fig. 9 . Choanocyte and archaeocyte after 2 hours incubation.
         For caption : see fig. 8.

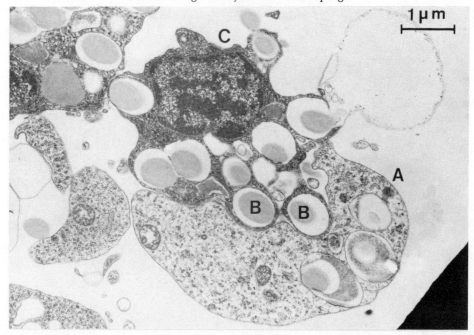

Fig. 10. Choanocyte and archaeocyte after 2 hours incubation. B = bead. A. = archaeocyte. C = choanocyte.

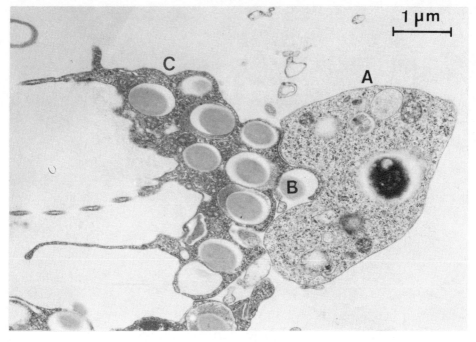

Fig. 11. Choanocyte and archaeocyte after 2 hours incubation. For caption : see fig. 10.

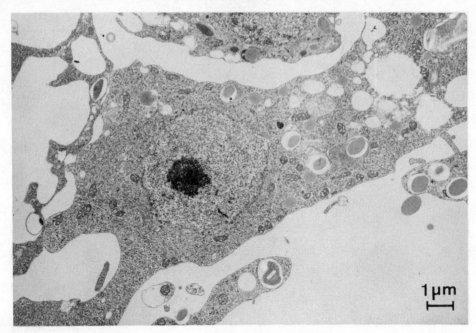

Fig. 12. Archaeocyte after 2 hours incubation.

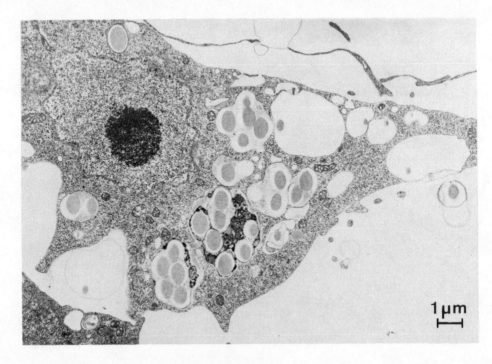

Fig. 13. Archaeocyte after 5 hours incubation.

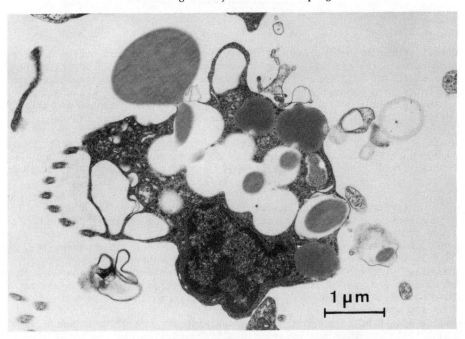

Fig. 14. Choanocyte releasing a bead into the mesohyl after
12 hours incubation.

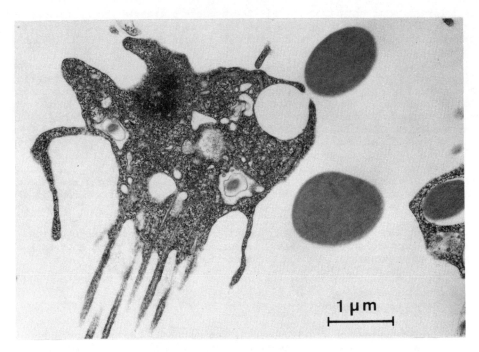

Fig. 15. Choanocyte releasing a bead into the mesohyl after
12 hours incubation.

DISCUSSION

The transit of particles ingested by *Ephydatia fluviatilis* has been studied from an ultrastructural point of view, using calibrated latex beads. Their ingestion timing by different cell types is listed in table 1. It clearly indicates that particle enter into sponges following two different pathways : the first and most common starting with particles phagocytosis by choanocytes, the second starting with particles phagocytosis by the exopinacocytes of the dermal membrane.

The particle retention by choanocytes is known since the first study on sponge feeding by Lieberkühn in 1856. It has been observed in electron microscopy by Fjerdingstad (1961), Pottu-Boumendil (1975), and Weissenfels (1976), but their further journey is still a point of controversy.

We have demonstrated that after having been phagocytosed by choanocytes, a lot of beads are directly transferred to archaeocytes lying by chance in close contact to the former. This confirms the results of Schmidt (1970) who observed the accumulation of fluorescent markers first in choanocytes and later in archaeocytes, but was not able to observe the cell to cell transfer (light microscopy).
Very recently, Diaz (1979) working on the brackish-water sponge *Suberites massa* claimed that choanocytes containing phagocytic vesicles do not transfer food to any other cell type but that they undergo a dedifferentiation, becoming themselves digesting archaeocytes.
Despite the fact that this hypothesis agrees with the out of date description of Masterman (1894) and with the reversible differentiation phenomenon of sponges (Reviewed by Rasmont, 1979), our results do not give any indication of such a deep morphological transformation of choanocytes.

It is interesting to note that not all the beads are transferred by cell to cell contact from choanocytes to archaeocytes, but that some of them are directly exocytosed into the mesohyl. The destination of such beads is still unknown.

Until today, the phagocytic ability of exopinacocytes was a point of controversy. Schmidt (1970) using fluorescent bacteria or baker's yeast as feeding material never observed phagocytosis of bacteria by *Ephydatia fluviatilis* exopinacocytes. She considered that if baker's yeast is exceptionally phagocytosed, this phenomenon must be considered as a cleaning function of the sponge surface, and not as a nutritional one. This point of view is shared by Reiswig (1971b) working on marine sponges *in situ*.

Our results indicate that phagocytosis of particles, (diameter of which is close to the size of bacteria) is very common at the level of the dermal membrane of *Ephydatia fluviatilis*. This leads us to assume that the dermal membrane is at least a second possible way of ingestion.

In conclusion, latex beads despite the fact that they are not a natural food appear to be particularly well suited tracers to study cell feeding exchanges in sponges. Their use have allowed us to demonstrate two possible pathways of ingestion, and to follow the destination of beads catched by choanocytes.

ACKNOWLEDGEMENTS

I would like to express my gratitude to G. Van de Vyver for her assistance in the preparation of this manuscript. Electron microscopy has been performed in the Laboratoire de Cytologie et d'Embryologie Moléculaires of P. Van Gansen; I am indebted to her for her cordial reception.

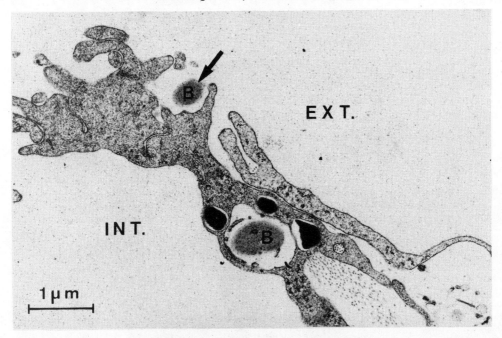

Fig. 16. Dermal membrane after 40 min. incubation.
Arrow indicates a bead undergoing phagocytosis.
EXT = external medium, INT : internal medium, B = bead.

TABLE 1. Timing of the bead ingestion by different cell types of *E. fluviatilis*.

Time		5 min.	10 min.	20 min.	30 min.	40 min.	1h.	2h.	5h.	12h.
Choanocytes			+	+	+	+	+	+	+	+
dermal membrane	Exo-pinacocytes			+	+	+	+	+	+	+
	collagen layer					+	+	+	+	+
	Endo-pinacocytes					+	+	+	+	+
Archaeocytes							+	+	+	+

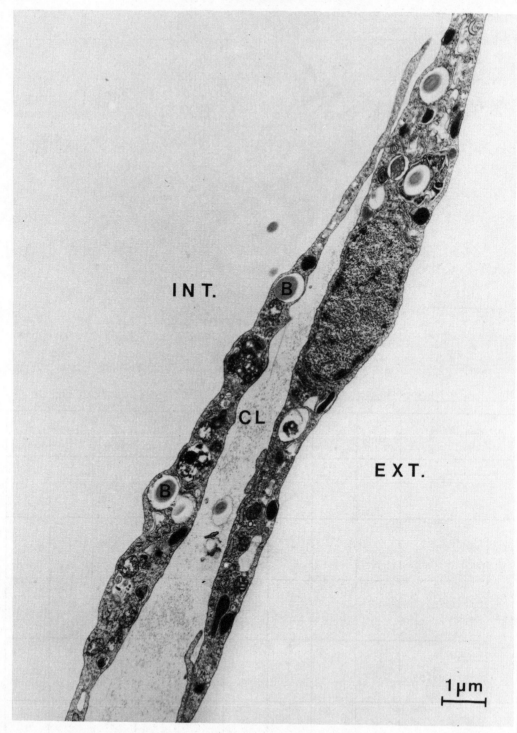

Fig. 17. Dermal membrane containing beads after 2 hours. B = bead.
EXT = external medium. INT = internal medium. CL = collagen
layer.

REFERENCES

Bagby, R.M. (1970). The fine structure of pinacocytes in the marine sponge
   Microciona prolifera (Ellis and Solander). *Z. Zellforsch. mikr. Anat.*,
   105, (4), 579-594.
De Vos, L. (1977). Morphogenesis of the collagenous shell of the gemmules of a
   fresh-water sponge *Ephydatia fluviatilis*. *Arch. Biol. (Bruxelles)*. 88, 479-494.
Diaz, J.P. (1979). Variations, différenciations et fonctions des catégories cellu-
   laires de la démosponge d'eaux saumâtres, *Suberites massa* Nardo, au cours du
   cycle biologique annuel et dans des conditions expérimentales. Thèse
   doct. Univ. Sc. et Tech. Languedoc, Acad. Montpellier, 1-332.
Fjerdingstad, E.J. (1961). The ultrastructure of choanocyte collars in *Spongilla
   lacustris L. Z. Zellforsch. Mikr. Anat.*, 53, 645-657.
Frost, T.M. (1976). Sponge feeding; A review with a discussion of some continuing
   research. In F.W. Harrison (Ed.), *Aspects of sponge biology*, Academic Press,
   New York pp. 283-298.
Harrison, F.W. (1972a). The nature and role of the basal pinacoderm of *Corvomeyenia
   carolinensis* Harrison (Porifera): Spongillidae). A histochemical and
   developmental study. *Hydrobiologia*, 39, 495-508.
Harrison, F.W. (1972b). Phase contrast photomicrography of cellular behaviour in
   spongillid porocytes (Porifera : Spongillidae). *Hydrobiologia*, 40, 513-517.
Kilian, E.F. (1952). Wasserströmung und Nahrungsaufnahme beim Süsswasserschwamm
   Ephydatia fluviatilis. Z. Vergleich. Physiol. 34, 407-447.
Lieberkühn, N. (1856). Beiträge zu entwicklungsgeschichte der Spongien. *Müller's
   Arch. f. Anat. u. Phys.*
Mastermann, A.T. (1894). On the nutritive and excretory processes in Porifera.
   Ann. Mag. nat. Hist., 13 (6), 485-496.
Pavans de Ceccaty, M. (1958). La mélanisation chez quelques éponges calcaires et
   siliceuses : ses rapports avec le système réticulo-histiocytaire. *Arch. Zool.
   expér. gén.*, 96, 1-52.
Pottu-Boumendil, J. (1975). Ultrastructure, cytochimie et comportements morphogé-
   nétiques des cellules de l'éponge *Ephydatia mülleri* (Lieb.) au cours de
   l'éclosion des gemmules. Thèse Doct. Spéci. Univ. Lyon, 1-101.
Pourbaix, N. (1931). Contribution à l'étude de la nutrition chez les Spongiaires
   (Eponges siliceuses).*Bull. Station Océnogr. Salambo*, 23, 1-19.
Pourbaix, N. (1932). Note sur la nutrition bactérienne des Eponges. *(Ann. Soc. roy.
   zool. Belg.*, 63, 11-15.
Pourbaix, N. (1933a). Mécanismes de la nutrition chez les Spongillidae. *Ann. Soc.
   roy. zool. Belg.*, 64, 11-20.
Pourbaix, N. (1933b). Recherches sur la nutrition des Spongiaires. *Inst. Español
   Oceanogr.*, 69 (2), 1-42.
Rasmont, R. (1961). Une technique de culture des éponges d'eau douce en milieu
   contrôlé. *Ann. Soc. roy. zool. Belg.*, 91, 147-156.
Rasmont, R. (1979). Les éponges : Métazoaires et sociétés de cellules. In *Biologie
   des spongiaires*. Coll. intern. CNRS, 291, pp. 21-30.
Reiswig, H.M. (1971a). *In situ* pumping activities of tropical Demospongiae.
   Mar. Biol. 9, 38-50.
Reiswig, H.M. (1971b). Particle feeding in natural populations of three marine
   Demospongiae. *Biol. Bull.* 141 (3), 568-591.
Reynolds, E.S. (1963). The use of lead citrate at high ph as an electron opaque
   stain in electron microscopy. *J. Cell Biol.* 17, 208-212.
Schmidt, I. (1970). Phagocytose et pinocytose chez les Spongillidae. Etude in vivo
   de l'ingestion de bactéries et de protéines marquées à l'aide d'un colorant
   fluorescent en lumière ultra-violette. *Z. vergl. Physiol.* 66, 398-420.
Simpson, T.L. (1963). The biology of the marine sponge *Microciona prolifera*
   (Ellis and Solander). I. A study of cellular function and differentiation.
   J. exp. Zool. 154 (1), 135-152.

Spurr, A.R. (1969). A low-viscosity epoxy resin embedding medium for electron microscopy. *J. Ultras. res.* 26, 31-43.

Van de Vyver, G. (1970). La non confluence intraspécifique chez les spongiaires et la notion d'individu. *Ann. Embr. Morph.* 3 (3), 251-262.

Van Weel, P.B. (1949). On the physiology of the tropical freshwater sponge *Spongilla proliferens* Annand. I. Ingestion, digestion, excretion. *Physiol. Comp. Oecol.*, 1, 110-128.

Van Tright, H. (1919). A contribution to the physiology of the freshwater sponges (Spongillidae). *Tijdschr. Ned. Dierk. Ver.*, 17, (2), 1-220.

Weissenfels, N. (1976). Bau und Funktion des Susswasser-Schwamms *Ephydatia fluviatilis* L. (Porifera) III. Nahrungsaufnahme, Verdauung und Defäkation. *Zoomorphologie* 85, 73-88.

Willenz, Ph. et R. Rasmont, (1979). Mise au point d'une technique de mesure de l'activité de filtration de jeunes éponges cultivées in vitro. In *Biologie des Spongiaires*. Coll. Intern. CNRS, 291, pp. 543-552.

INDEX

The page numbers refer to the first page of the article in which the index term appears.